T0327607

Introduction to DRYING of CERAMICS

Introduction to

DRYING of CERAMICS

With Laboratory Exercises

Denis A. Brosnan
and
Gilbert C. Robinson

The American Ceramic Society
735 Ceramic Place
Westerville, Ohio 43081
www.ceramics.org

ISBN: 1-57498-046-7

Cover Photos:
#1 Dryer at General Shale Brick (Moncure, N.C.). (photo courtesy of General Shale Brick)
#2 & 5: Fast dryer by Lingl Corporation at Rapis (Schwabmünchen, Germany).
#3 & 4: Commercial RF dryer.

Library of Congress Cataloging-in-Publication Data

Brosnan, Denis A,
Introduction to drying of ceramics: with laboratory exercises/Denis A. Brosnan and Gilbert C. Robinson.
p. cm.
Includes bibliographical references and index.
ISBN 1-57498-046-7 (alk. paper)
1. Ceramics—Drying. 2. Ceramics—Drying—Laboratory manuals. I. Title: Drying of ceramics. II. Robinson, Gilbert C. (Gilbert Chase), 1919- III. Title.

TP815.B77 2003
666—dc21

2003048019

For information on ordering titles published by The American Ceramic Society, or to request a publications catalog, please call 614-794-5890, or visit www.ceramics.org.

Dedication

Gilbert C. Robinson

"And they said one to another, let us go make brick and burn them thoroughly."
—Genesis 11:3

Dedicated to the Robinson family:
Barbara, Barbie, Chandler, Chase, and Tim

With loving appreciation to the Brosnan family:
Bev, Beth, Denis, and Rob

Contents

CHAPTER 2: WATER, AIR, AND WATER VAPOR

CHAPTER 3: DRYING MECHANISMS IN PARTICULATE SYSTEMS

CHAPTER 8: ADVANCED DRYING TECHNOLOGIES

CHAPTER 9: LABORATORY EXERCISES

List of Figures

List of Tables

Preface

Gil Robinson was working on the manuscript of his book on drying prior to his death in February 1996. He and I had talked about writing a book on drying, but we never formally started the book together. After his death, I accepted the job of completing his book to ensure that this important work would proceed to publication. I have not attempted to change the substance of his writing, but I have changed the form considerably and have added course materials and problems from my own university courses.

Prior to his passing, Dr. Robinson was working on the chapter on dryer control with Don Denison, and Mr. Denison has graciously completed that chapter. John Sanders and I have written a chapter on advanced drying systems. Mr. Denison, Dr. Sanders, and I were students of Dr. Robinson.

In order to accommodate readers with a range of interests and abilities, most chapters are divided into sections by basic and advanced concepts. The sections on basic concepts are intended for those readers who are interested in the practical issues and usually involve only simple mathematics. The sections on advanced concepts involve more theoretical analysis using appropriate mathematical expressions.

The terminology used in this book is similar to that in the *Chemical Engineer's Handbook* (John Wiley and Sons). This terminology was adopted to make the book consistent with that of the widest usage in engineering practice. English dimensions or units are used consistent with current engineering practice in North America, and metric dimensions are also provided in most sections. Text boxes indicate important facts or key concepts for readers.

This book is one form of the legacy of Gil Robinson's life and his contributions to the body of ceramic engineering technology. This service spanned over 50 years, including his service in the United States Navy in World War II. When asked why he chose ceramic engineering, he would simply relate that it seemed like an expanding field that yielded products needed by society. Whatever the reason, the ceramic engineering field benefited from his decision to attend North Carolina State University and graduate with a degree in ceramic engineering.

Dr. Robinson's accomplishments as an educator are astonishing. He came to Clemson University in 1946 as state and university leaders sought to exploit mineral resources. He started with very little in terms of facilities or equipment, yet he was able to teach courses to a group of young people destined to become industry leaders. In retrospect, the Olin Foundation's gift to provide an academic building to house the ceramic engineering department was a miracle of its day.

While his service as department head for some 40 years is a sign of his dedication, the length of his service tells only part of the story. Gil Robinson was "father" and "banker" to a large group of early students, as well as professor and mentor. There are numerous stories of young families who couldn't buy groceries or pay tuition, and Dr. Robinson reached into his pocket to keep his students going. In later years, Gil Robinson was "quality control" in the department. He earned his nickname: "the Chief."

Gil Robinson's success as an educator is evidenced by the success of his graduates in traditional ceramic industries. Clemson's reputation in these fields, especially the brick industry, is such that its graduates are know for being able to "function on their feet" in processing jobs in industry. This reputation is largely attributable to Gil Robinson's personal efforts as an educator. Clemson University recognized his efforts recently by renaming the department "The Gilbert C. Robinson Department of Ceramic Engineering" in the early 1990s.

The legacy of Gil Robinson lives on with the publication of this book. This is a rich legacy of improvements in processing in traditional ceramic materials and effective education. Gil Robinson was many things to many people, including teacher, administrator, author, innovator, mentor, husband, father, and friend. It is a life worthy of celebration.

Heartfelt thanks are expressed to Teresa Williams of Clemson University for her assistance with the manuscript. Ms. Williams worked with Dr. Robinson over a period exceeding ten years.

I must also recognize the support of my wife, Bev, and our children, Denis Jr., Rob, and Beth, during the six long years this book took to prepare, with most holidays devoted to writing. In addition, I want to thank some significant friends for their encouragement over the years. Tom Keinath has been an example of perseverance, dignity, and wisdom in management, and his oversight has been a key to the success of the National Brick Research Center. Chris Przirembel has been a trouper, significantly helping with the industry–university partnership of the Center. I shall remain ever grateful to Paul Hummer for his formative influence and dignified example in early years. Blair Savage has been an example in the fair-minded treatment of individuals, and he started me on the road to higher levels of service.

The influence and generosity of the late George J. Bishop III, his wife, Dot Bishop, and his son, David Bishop, all long-time friends of Gil and Barbara Robinson, benefactors to the National Brick Research Center, are worthy of special mention. I also thank the executive committee of the National Brick Research Center for providing time for me to complete the manuscript. Particular recognition is appropriate for the friendships of Pete Cieslak, J. B. Cooper, Bill Kjorlien, John Koch, Terry Schimmel, and Mitch Wells. This book would have been impossible without the professional assistance of Jim Frederic.

I shall remain ever grateful to Gil Robinson, my wife, and all of my "bosses" named above.

Denis A. Brosnan
Clemson, South Carolina
August 2002

CHAPTER 1

Introduction to Drying

PART A: BASIC CONCEPTS

1.1 Introduction

A wise old professor asks a student to define the variables in drying ceramic products — a likely question for any student interested in ceramic drying processes. Even if the student is not familiar with the techniques of ceramic drying, he or she can rely on everyday experiences to answer the question because drying processes are a part of our everyday lives. From drying clothes to drying our dishes, we understand that the air temperature and the air's ambient humidity play a part in the rate of drying. How wet the objects are, that is, the quantity of water to be evaporated, determines to a large degree how fast drying can be accomplished. The airflow available to accelerate drying is an obvious factor in determining the overall rate of drying.

However, drying of ceramics is much more complicated than drying many other objects because unfired ceramics typically exhibit shrinkage during drying. This shrinkage can lead to cracking and loss of acceptable quality in production. Another complicating factor is that ceramic dryers are typically very large machines.

Drying is the process of removing water from an unfired ceramic object or raw material in the green or as-formed state or in the as-received state. As such, drying is accomplished by supplying energy to the ceramic in order to accomplish evaporation. In some cases, mechanical methods such as filtration are used to reduce the moisture content of ceramic raw materials (for example, ceramic slurries). However, the term *drying* in ceramic processing is restricted to the process of reducing moisture content of as-formed ceramic parts or powders using application of energy. Drying is continued to a lower moisture content at which the ceramic

DRYING IS A PROCESS:

- Used to remove water at a rate compatible with the overall process rate.
- Conducted at an acceptable cost to the producer.
- Conducted in such a way as to not damage the ceramic.

can be subsequently *fired* or *sintered* at temperatures exceeding about 900°C ("red heat"). Fired ceramic objects can also be dried prior to further processing, such as decoration, or prior to high-temperature exposure to ensure that steam spallation will not occur.

The goal of drying is to remove water at a rate required by the overall process, at an acceptable cost, and in such as way as to not damage the ceramic so that fired physical properties are at optimum levels. Ceramic products are formed by mixing particulate matter and additives with a liquid to provide plasticity and cohesion in forming. Sufficient cohesion is necessary so that the formed object may be moved into ovens to accomplish drying.

Since the liquid used in forming is usually water, the discussion in this book will concern primarily evaporation of water. However, drying ceramics containing organic liquids is analogous to drying ceramics containing water — meaning the same principles are involved. One important difference is that drying substances containing organic liquids can involve handling of organic vapor–air mixtures that may be flammable or explosive.

1.2 Expressing the Moisture Content of a Powder or a Ceramic Product

Ceramic products are made from mixtures or *batches* of particulate matter and a liquid. Since this liquid is usually water, it is important to express the moisture content in a way that there can be no confusion as to how it was determined. In European practice, the moisture content may be called the *humidity* of the product. The two ways to express moisture content are *dry weight basis* and *wet weight basis.*

APPROXIMATE RANGE IN WATER CONTENTS IN AS-FORMED CERAMIC PRODUCTS, (% DRY WEIGHT BASIS):

Isostatically pressed Technical ceramic	<0.1+%
Dry-pressed refractory	<0.5% to ~1.5%
Dry-pressed tile	~0.1–0.8%
Extruded brick	~14–18%
Molded brick	~22–25%
Tableware (jiggered)	~9.0–16%
Sanitaryware	<12–16%+
Engobes	~16–20+%

Ceramic products as-formed and prior to drying are referred to as *green* or *in the green state.* The moisture content is expressed in word equations as follows:

$$\text{Moisture (\% dry basis)} = \frac{\text{Weight of water in the specimen}}{\text{Dry weight of the ceramic or powder}} \times 100 \quad \textbf{(1.1)}$$

or

$$\text{Moisture (\% wet basis)} = \frac{\text{Weight of water in the specimen}}{\text{Dry weight of the ceramic or powder} + \text{weight of water}} \times 100 \quad \textbf{(1.2)}$$

The latter equation for moisture on a wet basis is the same as:

$$\text{Moisture (\% wet basis)} = \frac{\text{Weight of water in the specimen}}{\text{Green weight of the ceramic specimen}} \times 100 \quad \textbf{(1.3)}$$

Some ceramic producers use a different weight basis than others in reporting mix compositions. The refractory and technical ceramic industries usually use a dry weight basis while the brick and sanitaryware industries usually use a wet weight basis. If the basis is not identified when reporting a composition, there is a chance that a batching error could be made, resulting in production of defective products.

It is the best practice to express the percentage of moisture in a ceramic with the accompanying designation *DB* or *WB* to indicate dry basis or wet basis, respectively. For example, a report of 3.0% moisture WB means that the material was found to contain 3.0% moisture on a wet weight basis. In some industries, it is not necessary to express DB or WB because one or the other is used by convention. If the basis is not known, serious errors could develop in processing, as illustrated in the example below.

EXAMPLE 1.1: A green ceramic batch is produced from a blend of 200 lb each of potters flint (–200 mesh ground silica), kaolin clay, and feldspar with 11 gal of water. The density of water is 8.34 lb/gal (~8.3 lb/gal). Express the batch composition on a dry basis and a wet basis:

Solution:

$$11 \text{ gal of water @ } 8.3 \text{ lb/gal} = 91.3 \text{ lb of water} \approx 91 \text{ lb of water}$$

$$\text{Moisture (dry basis)} = \frac{91 \text{ lb water}}{600 \text{ lb powder}} \times 100 = 15.2\%$$

$$\text{Moisture (wet basis)} = \frac{91 \text{ lb water}}{600 \text{ lb powder} + 91 \text{ lb water}} \times 100 = 13.2\%$$

1.3 Adding Water to a Completely Dry Ceramic and Subsequently Removing that Moisture

When water is added to a completely dry ceramic powder, the initial portion of water will cover the surface of the particles, assuming that ideal mixing is achieved (this water is usually called the *surface water* or the *hygroscopic water*). The phenomena in this initial coverage of the surface involve the electrical nature of the surface and the polar nature of the water molecule (discussed in Chapter 2).

There is an unbalance of electrical forces at the surface of a liquid and a solid resulting from incomplete or noncontinuous bonds at the interface. When water is added to a ceramic powder, the water generally "sees" a negative charge cloud at the ceramic surface. In order to minimize the energy of the surface, the ceramic surface attracts the positive side of the polar water molecule toward the surface to achieve a degree of wetting. Van der Waals forces develop, holding the water molecules onto or in close proximity to the surface.

Many ceramic powders exhibit a consistent attraction for water, a characteristic known as the *moisture adsorption* property of the powder; that is, an adsorption equilibrium is attained between a powder and an atmosphere of fixed humidity. The moisture adsorption quantity is related to the nature of the ceramic surface and the exposed surface area as well as to the percentage of relative humidity in the atmosphere. Since surface area depends on particle size distribution, the moisture adsorption is strongly dependent on the particle size distribution or *fineness* of the ceramic powder.

Table 1.1: Typical moisture adsorption of selected ceramic raw materials

(20°C at >40% relative humidity)

Material	Moisture Absorption, %
Calcined alumina	<0.3
China clay (kaolin)	1.0–1.5
Brick clay and fireclay	0.5–4.0
Ball clay	5.5–7.5
Bentonite	7.5–14.5

Ceramic materials generally increase in moisture adsorption near room temperature as relative humidity increases. Moisture adsorption for ceramic powders typically reaches a maximum beyond ~30–40% relative humidity. Some approximate equilibrium moisture contents for ceramic materials are shown in Table 1.1.

WATER ADDED TO A DRY CERAMIC POWDER:

- First, covers the surface of the particles. (assuming ideal mixing to prevent agglomeration).
- Second, fills small capillaries (space between particles).
- Third, fills pores.
- Fourth, causes a separation of particles.

DRYING REMOVES:

- Forming water first (if present).
- Pore water next (if present).
- Capillary water (usually present).
- Surface water (always present).

After the surface adsorption requirements for water are met — that is, when the water covers the surface in thin molecular layers — the water next fills voids between fine particles known as capillaries (called the *capillary water*). Capillaries are generally defined as voids or pores of a diameter less than about 1 μm in diameter or size, and since most ceramic mixes contain micrometer-sized powders, submicrometer voids are usually present.

At the same time, continuous water films are created around particles, giving rise to an attraction between particles via surface tension due to overlap of water films between particles (traditionally called *Norton's theory of plasticity*). This is the point at which agglomeration is noted in low-intensity mixing as water is added to a dry ceramic powder; that is, *balling* or *agglomeration* of the mix begins. Agglomeration occurs because water drops or segments cause a high moisture content within a limited volume of the batch being mixed. More intense mixing or long mixing times are required to fully distribute the water.

Eventually the water is distributed, and the mix no longer exhibits dust generation on agitation. Since most ceramic compositions also contain some larger pores (generally ~50–100 μm in size) due to imperfect particle packing or gradation, these pores are gradually filled with water. In mixes that exhibit plasticity, the mixer must work harder to achieve water distribution during filling of the larger pores. This water quantity, over and above surface and capillary moisture, is called the *pore water.*

Finally, additional water tends to significantly separate particles in the material. In plastic forming processes, this separation adds lubricity to the system to facilitate forming. Also, the mixer must work even harder, and a point of maximum apparent viscosity is reached. Measurements of energy in mixing are, in fact, used to quantify plasticity of materials. This last water quantity that causes separation of particles is called the *forming water.* Some practitioners use the term *pore water* to include both the water filling large voids and the water employed in separating particles.

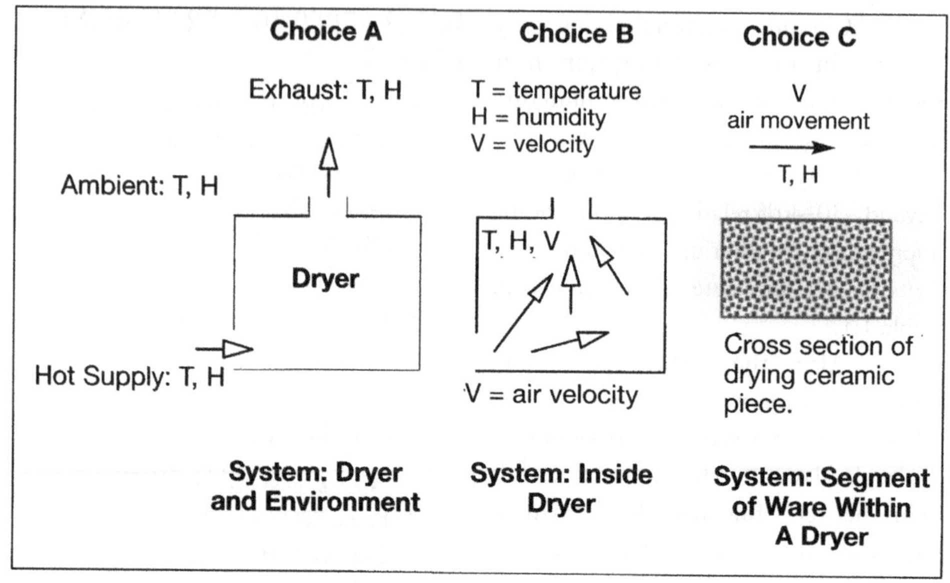

Figure 1.1: Choices of the system in drying problems.

In drying, water is removed from the ceramic, and the sequence in which the various waters are removed is exactly the reverse of the theoretical sequence of water addition given above; that is, first forming water, then pore water, then capillary water, and finally surface waters are removed.

If the ceramic was made by a process in which all forms of water are not present (as in dry pressing), the order is the same starting with the last form of water created when the mix was made. Particulate and water interaction is discussed in subsequent chapters.

1.4 Systems and Segments of Space

In order to discuss drying phenomena it is necessary to discuss events in a segment of space conventionally called the *system of interest.* With respect to drying, there are several choices depending on the perspective of the observer.

The system can be chosen as the dryer and its immediately surrounding space, it can be chosen as the volume within the dryer, or it can be chosen as a segment within the drying product (Figure 1.1). There is no magic to the choice of the system. It is just a matter of convenience for the discussion. However, the definition of the system is important in understanding application of drying principles.

In discussing the rate or mechanism of moisture movement within the ceramic during drying, the system may be defined as a limited volume element within the particulate system or as the formed piece (Choice C in Fig. 1.1). When there is concern with balancing the thermal conditions and the air movement within the dryer, the system may be defined as the interior or inside of the dryer (Choice B in Fig. 1.1). For determining the thermal efficiency of the dryer, the whole dryer and the environment are typically chosen as the system of interest (Choice A in Fig. 1.1). So the choice of the system is simply a matter of convenience.

1.5 Evaporation and Convection Drying Processes

Drying of ceramics is accomplished by the evaporation of water, that is, the change of state of water from a liquid to a gas in the form of water vapor. Thermal energy is required to accomplish this change of state. This thermal energy is called the *latent heat of vaporization,* that is, the amount of thermal energy required to accomplish the phase change without any change in temperature (Table 1.2).

The term *latent* is not intended to be confusing. It simply means that there is heat (energy) that can be available or recovered when desired. It is similar in concept to potential energy. For example, the energy expended in vaporizing water is "carried" with the water vapor to be recovered when the water vapor is condensed. This is analogous to the potential energy of an elevated object before it is dropped — kinetic energy develops under the acceleration of gravity when the object is released.

Heat is energy moving from a high-temperature source to a receiver. Heat is measured in joules (J), where 1 J is the quantity of heat required to increase the temperature of 1 g of air from 100 to 101 K. The mechanical equivalent of 1 J is 1 N of force applied over a distance of 1 m. In the English system of units, heat is measured in terms of the *British thermal unit* (Btu), defined as 1/180 of the energy required to increase the temperature of water from 32 to 212°F.

It is convenient to define the ***Btu*** as the quantity of energy required to raise the temperature of 1 lb of water by 1°F and the ***kilocalorie*** as the quantity of energy required to raise the temperature of 1 kg of water by 1°C.

It takes much less energy to increase the temperature of water by 1 K (1 Btu/lb or 4.18 J/g) than to evaporate the same quantity of water (973 Btu/lb or 2263 J/g). *This is the primary reason that a dryer must effectively transfer heat into the ceramic to accomplish drying.*

Table 1.2: Latent heat of vaporization of water at 212°F (100°C) at 1 atm pressure

9729 cal/mol
2263 J/g
540.5 cal/g
973 Btu/lb

The fundamental reason for the high latent heat of vaporization of water

Table 1.3: Specific heats of selected materials at 86°F (30°C)

Material	Specific heat (cal/g/°C)
Water	1.00
Water vapor	0.46
Dry air	0.24
Clay	0.22
Alumina	0.19
Quartz	0.21
Magnesia	0.23
Concrete	0.19
Graphite	0.31
Glass	0.20
Iron	0.11
Gold	0.03
Teflon	0.24
Bakelite	0.33

is hydrogen bonding in the liquid phase, which retards evaporation until sufficient energy is available for water molecules to escape into the vapor phase. The subject of hydrogen bonding is discussed in Chapter 2.

The quantity of heat required to raise the temperature of 1 g of a substance by 1°C of temperature is called the *specific heat* of that substance. As shown in Table 1.3, the specific heat of water and water vapor far exceed the specific heats of most solids. Specific heats of materials increase with temperature, but this increase is negligible in the temperature range of interest in drying and thus usually can be ignored.

Heat capacity is the quantity of heat required to increase the temperature of a molar quantity of a substance by 1°. The value of heat capacity for most ceramic materials approaches a value of 24.9 J/mol above 1000°C. The definition of a *mole* is the familiar *gram atomic weight* or *formula weight* of traditional chemistry.

Another measure of energy in engineering or scientific use is the calorie (cal), defined as 1/100 of the energy required to increase the temperature of 1 g of water from 0 to 100°C. Since the calorie is a relatively small unit of measure, the kilocalorie (kcal) is usually employed. In popular terminology, the energy from consumption of food or expended in exercise is conventionally called "calories," but that same energy unit in diet or exercise is actually the kilocalorie of engineering practice. In all cases, the terms *calorie* and *kilocalorie* will be used in this book per the engineering or scientific definition.

It is convenient to define the ***calorie*** as the quantity of energy required to raise the temperature of 1 g of water by 1°C.

The rate of energy transfer in a dryer is governed by the principles of heat transfer and aerodynamics. The three mechanisms or modes of heat transfer are:

- *Conduction* (energy transfer through a solid or through solids contacting each other).
- *Convection* (energy transfer from a gas phase to a solid object or another air segment or vice versa).
- *Radiation* (energy transfer by electromagnetic waves from one segment of space to an object or another segment of space or vice versa).

To illustrate the modes of heat transfer in drying, consider a 100 mL capacity beaker of water filled with 100 g of water initially at 100°C (Figure 1.2). If the beaker is placed on a laboratory hot plate whose surface temperature exceeds 100°C, heat will be transferred by conduction from the hot plate surface through the beaker into the water, eventually causing the water to boil (evaporation of all the water). If, however, the beaker is placed instead on a counter top with a fan blowing hot air at 100°C onto the beaker, heat would be transferred by convection (transfer of energy by air molecules contacting the beaker and exposed water surface, eventually evaporating all of the water). Another way of evaporating the water in the beaker is to place it on the counter top under an infrared lamp until evaporation is complete by a radiation

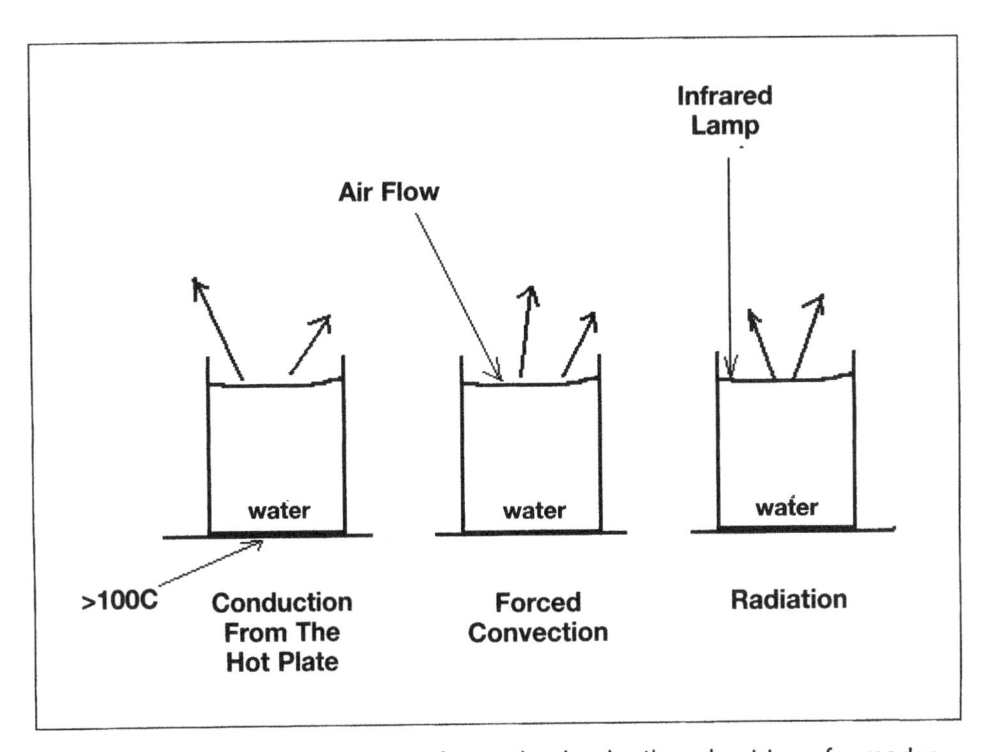

Figure 1.2: Evaporation of water from a beaker by three heat transfer modes.

heat transfer mode. In all three cases, even though the heat transfer mechanism is different, the energy consumed in evaporating the water was the same (226,300 J for 100 g or 540,500 cal for 100 g). In fact, attaining a temperature of 100°C is not necessary for causing evaporation if the relative humidity above the beaker is less than 100%.

In actual practice, more than one mode of heat transfer is taking place in any given drying situation. However, one mode of heat transfer is usually predominant. In most large dryers, convection is the predominant mechanism of heat transfer; that is, hot air is used to heat the ceramic being dried in order to achieve the dryness required in the process. The air also serves as a vehicle to convey water vapor away from the drying ceramic, and the moisture content of the drying air is one of the determinants of the rate of evaporation.

Dryers employing forced airflow for heating and removal of water vapor are called *convection dryers.* They may operate in *batch mode,* drying one production unit of product per drying cycle, or in *continuous mode,* in which sequential batches of product are inserted into and correspondingly extracted from the dryer as shown in Fig. 1.3. The continuous *countercurrent* convection dryer is so named because the product and airflow are in opposite directions; this type of dryer currently predominates in most tonnage-scale ceramic processing applications.

1.6 Enthalpy, Energy Balances, and Reference Temperatures

Enthalpy is the term used to describe the energy content of a substance that can be *exchanged* or recovered in a process resulting in a change in temperature in the substances involved in the exchange. From another perspective, it is a measure of the quantity of heat required for a change in the temperature of a substance at constant pressure. Since most drying processes are at 1 atm of total pressure, drying processes involving total pressure changes are not considered in this text.

Enthalpy is simply "recoverable" or "exchangeable" energy.

Enthalpy content is a quantity relative to a reference point known as the *standard state.* Since we observe enthalpy changes rather than measure absolute enthalpy content, we do not need to know the absolute amount of energy at any point defining the state of a substance. In the simplest case of a material heated at constant pressure, the incremental enthalpy change of that substance is given by the equation:

$$dh = mC_p dT \tag{1.4}$$

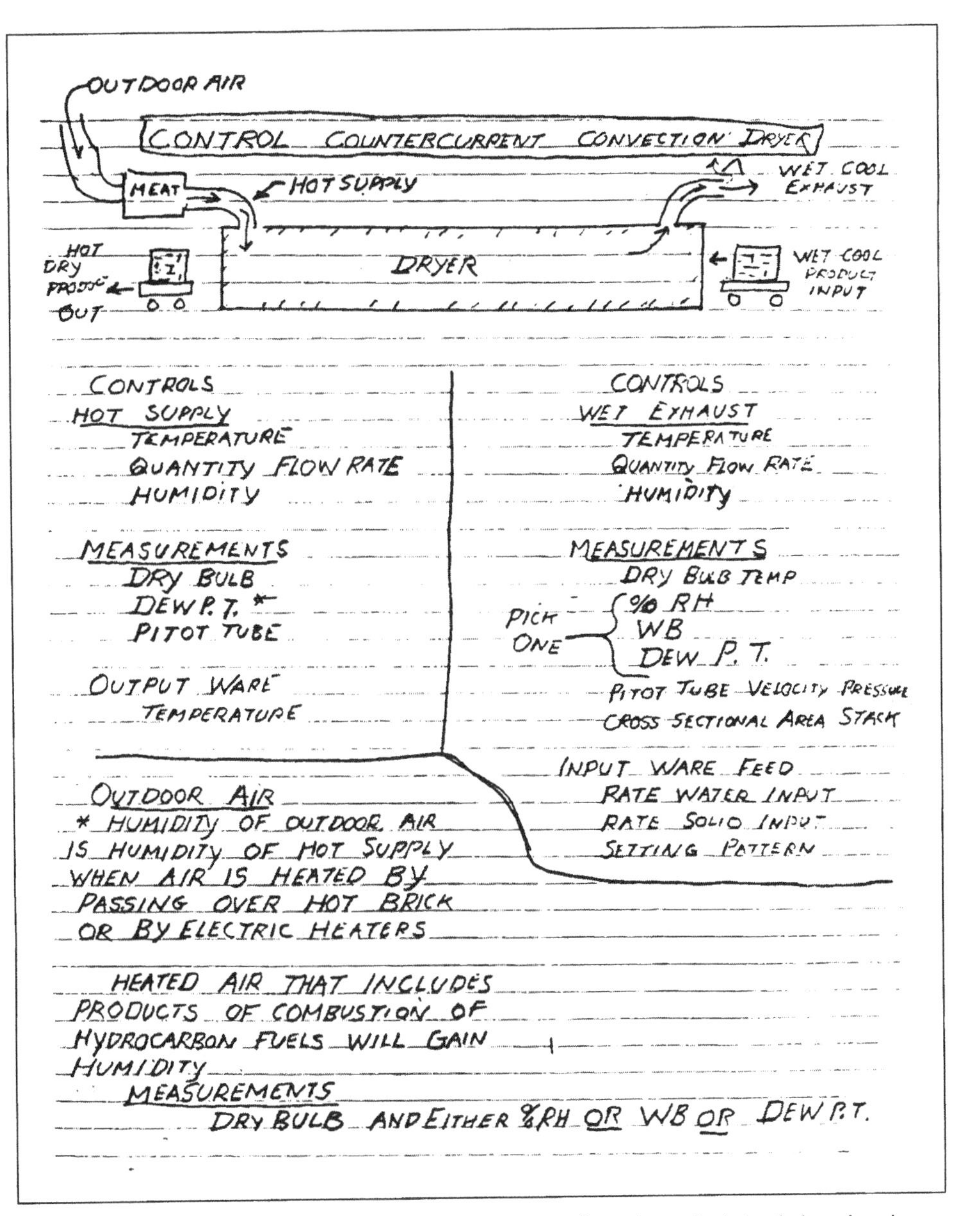

Figure 1.3: Continuous countercurrent convection dryer (original drawing by Gil Robinson).

where dh is the incremental increase in enthalpy simply due to heating, m is the mass of substance being heated, C_p is the heat capacity of the substance at constant pressure, and dT is the temperature change involved in changing the substance from Temperature 1 (the lower temperature) to Temperature 2 (the higher temperature).

The equation above can be stated more simply as follows:

$$\text{Enthalpy change } (\Delta h) = \text{mass} \times \text{specific heat} \times \text{temperature change} \quad \textbf{(1.5)}$$

or

$$h = (m)(C_p)(\Delta T) = (m)(C_p)(T_1 - T_2) \quad \textbf{(1.6)}$$

where the symbol Δ is used to denote change in the quantity of interest, and T_2 and T_1 are the higher and lower temperatures of interest, respectively.

The change in enthalpy can be more complex, and the generalized formula for n components can be applied:

$$\Delta h = m_1 C_{p,1} dT + m_2 C_{p,2} dT + \ldots + m_n C_{p,n} dT + h_{\text{transitions}} + \cdots \quad \textbf{(1.7)}$$

where the quantity $\Delta h_{\text{transitions}}$ is applied in the event that phase changes or other transitions take place. This simply means that the enthalpy change is due to the recoverable energy changes in the material, whatever they may be (vaporization, solidification, etc.).

EXAMPLE 1.2: What is the enthalpy change on heating from 0 to 100°C of a 100 g green ceramic part consisting of aluminum oxide powder molded at 3% water (WB)? Assume no evaporation of water until 100°C is reached (for this example only).

Solution:

The part is constituted of 3 g of water and 97 g of alumina powder. The specific heats from Table 1.3 are 1 cal/g water and 0.19 cal/g alumina. Therefore:

$$\Delta h = \text{energy consumed in heating water} + \text{energy consumed in heating alumina}$$

From Eq. 1.7:

$$\begin{aligned}\Delta h &= (3\text{ g})(1\text{ cal/g°C})(100\text{°C}) + (97\text{ g})(0.19\text{ cal/g°C})(100\text{°C}) \\ &= 300\text{ cal} + 1843\text{ cal} \\ &= 2143\text{ cal}\end{aligned}$$

The energy content of the ceramic part has been increased by 2143 cal by transfer of this energy from the environment into the part. Note that it is possible to complete this problem using English units or dimensions.

Enthalpy changes can be due to changes involving *sensible* heat and or *latent* heat. Sensible heat or *sensible enthalpy* changes are those that result from a temperature change but without any change in state of a component in the system (e.g., there is no evaporation). With respect to drying, sensible heat changes are those required to elevate the temperature of a mass of air and water vapor without any change in the ratio of water vapor to air in the system. Sensible heat changes are those that can be sensed or measured with a temperature-measuring instrument (such as a thermometer).

Latent heat or *latent enthalpy* changes are energy changes due to the introduction of water vapor into the system and energy changes due to water evaporation. As discussed earlier, the basic concept is that if water is evaporated into an air mass, the water carries with it the amount of energy required to cause the evaporation (i.e., the latent heat), plus it carries the energy originally used to heat the air–water mixture. On cooling the air mass to the point at which condensation occurs, the latent heat can be recovered in the system. The latent heat of vaporization of water is large (Table 1.2).

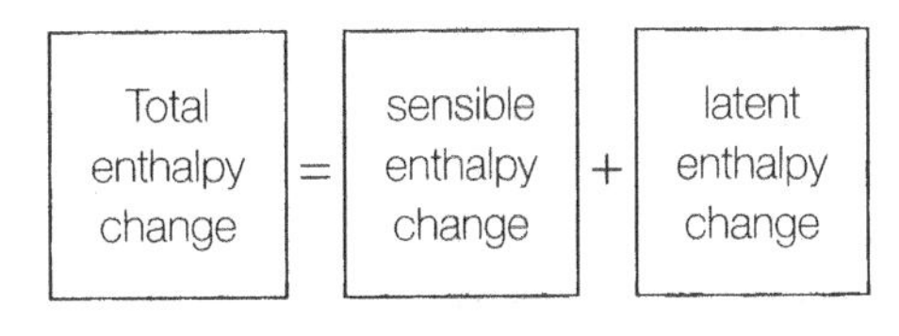

It turns out that water vapor increases the ability of air to carry or contain energy. The role of the presence of water vapor can be illustrated by considering the example below.

EXAMPLE 1.3: The enthalpy change of 1 g of dry air on heating from 20 to 100°C is compared to that of heating 1 g of a mixture of 99% dry air and 1% water vapor by weight through the same temperature range is:

For dry air:

$$\Delta h = mC_p\,dT = (1\text{ g})\,(0.24\text{ cal/g})\,(80°\text{C}) = 19.2\text{ cal}$$

For the mixture of 0.99 g of dry air and 0.01 g of water vapor (using data from Table 1.3):

$$\begin{aligned}\Delta h &= \text{enthalpy change of air} + \text{enthalpy change of water vapor}\\ &= (0.99\text{ g})\,(0.24\text{ cal/g})\,(80°\text{C}) + (0.01\text{g})\,(0.46\text{ cal/g})\,(80°\text{C})\\ &= 19.4\text{ cal}\end{aligned}$$

Conclusion: One gram of humid air (mixture of 99% air and 1% water vapor) contains about 1% more energy [(19.4 – 19.2) / 19.2] × 100 than 1 g of dry air on heating from 20 to 100°C. This is a "real" problem since air is capable of holding up to about 1.6 g of water per gram of dry air at 20°C (discussed in Chapter 4). Note that this increase in energy was simply due to an increase in sensible enthalpy content.

Now consider the enthalpy change of 1 g of dry air heating through the same interval (20 to 100°C) when 1% liquid (water) is added to the air along with the energy necessary to vaporize the water. In this case the water will be vaporized during heating. What will be the enthalpy content of the mixture at 100°C?

$$\begin{aligned}\Delta h &= \text{enthalpy change of air} \\ &\quad + \text{enthalpy of vaporization of the water} \\ &\quad + \text{enthalpy change of water vapor} \\ &= (0.99\text{ g})\,(0.24\text{ cal/g})\,(80^\circ\text{C}) + (0.01\text{ g})\,(540.5\text{ cal/g}) \\ &\quad + (0.01\text{ g})\,(0.46\text{ cal/g})\,(80^\circ\text{C}) \\ &= 24.8\text{ cal}\end{aligned}$$

The enthalpy content of the air has been increased significantly by adding latent heat to the total energy change, that is, the total increase in sensible and latent heat content of the air segment.

A typical question is whether the actual temperature of evaporation affects the enthalpy content of the humid atmosphere. The answer is that it does not matter (discussed in Part B of this chapter).

The above example is important because it illustrates the role of water vapor in increasing the energy content of humid air. What is more important from the perspective of drying is that the energy can be recovered or removed from the systems by cooling the atmosphere back to a desired temperature, for example, 20°C. Here we are letting the atmosphere do the work; that is, transferring heat into some component of the system is analogous to letting air do work in a dryer by transferring heat into a product to be dried. The recoverable heat from the above example is summarized in Table 1.4.

Convection dryers use relatively hot and dry air to cause drying to occur in relatively cooler and moist ceramic materials. Thereby, energy exchanges are occurring during drying between the air and the ceramic, leading to the exchange of mass (i.e., evaporation of water), which in turn changes the energy content of the atmosphere. These exchanges involve changes in both sensible and latent heat in the air moving through the dryer. Since the air is transferring energy to the ceramic, the dryer air loses

Table 1.4: Maximum recoverable energy of humid air segments on cooling

Original composition of 1 g segment	Enthalpy recoverable (100°C → 20°C)
100% dry air	19.2 cal
99.0% dry air + 1.0% water vapor	19.4 cal
98.4% dry air + 1.6% water vapor (saturated air at 20°C)	19.5 cal*
99.0% dry air + 1.0% water (condensed)	24.8 cal*

**Not shown in the example above but easily calculated*

or decreases in sensible heat. However, the energy transferred causes evaporation, and the water vapor released into the dryer air (or *draft*) carries energy with it in the form of latent heat.

It is convenient to assume *adiabatic* conditions in drying problems. An adiabatic change is one in which there are no exchanges of energy across the boundary of the system of interest. An air segment may be considered to be adiabatic if it can exchange energy only within its own segment of space. Within an adiabatic air segment, the loss in sensible heat equals the increase in latent heat during drying. In this sense, the dryer is exchanging sensible heat for latent heat among air segments to accomplish drying.

Enthalpy changes are based on arbitrarily chosen starting *reference temperatures*. For scientific purposes, reference temperatures are usually chosen as the freezing or boiling point of water. In drying, it is convenient to relate to ambient conditions, such as the air used as *feedstock* to the dryer inlet, and it is customary to choose outside or ambient plant air as the reference point. Even in dryers fed with waste heat from a kiln, outside air conditions are usually chosen as the reference point (reference enthalpy).

Enthalpy changes in drying involve exchanges between sensible and latent heat. Adiabatic changes involve no losses or gains across the "boundary" of the system.

1.7 Trends in Drying and Advanced Drying Processes

In recent years, *fast drying* has become important in those countries where energy restrictions are imposed by regulators and/or where energy costs are high. Work with fast drying has generally involved achieving maximum use of the air available for convection drying by improving circulation within the dryer.

Pulse or *rhythmic drying* has been investigated for several years. In this method, a momentary blast of drying air is directed to the part's surface followed by a

ADVANCED DRYING PROCESSES:

- Fast drying
- Pulse or rhythmic drying
- Dielectric drying
- Radio frequency drying
- Microwave drying

momentary respite of no air impingement, with continuous on and off cycling. This theoretically allows a dwell period for reduction of surface stresses, particularly in the initial stage of drying when high shrinkage can be observed. Proponents of rhythmic drying claim that lower scrap rates may be achieved in these processes.

There has also been recent work in high-frequency electric heating of ceramics to accomplish drying, such as *dielectric drying, radio frequency drying,* and *microwave drying.* These approaches use imposed electric fields on the green ceramic that change direction, causing water molecules to rotate against frictional constraints and resulting in internal heating in the green ceramic part.

The dielectric drying processes (radio frequency and microwave drying) provide for more rapid drying because the rate of heat transfer into the part is much higher than in convection dryers. In addition, the part may exhibit a smaller shrinkage gradient, particularly in parts with hollow spaces or cores, resulting in higher yields of acceptable-quality parts.

1.8 Drying Products Containing Nonaqueous Liquids

Nonaqueous liquids used in ceramics include substances added as binders, plasticizers, and lubricants. Some of these include fuel oils, pitch and resins, polyethylene glycol, and alcohol. Many of these may polymerize and yield a carbon char if heated rapidly; the drying strategy may be to remove as much binder as possible by evaporation or depolymerization processes prior to rapid heating. The consequences of residual carbon within the material can include formation of chemically reduced species in firing and reduced wetting upon vitrification, usually resulting in an undesired change in key physical properties. One well-known consequence of residual carbon during firing is the formation of *black cores,* which, if accompanied by bloating, results in a loss in strength of the fired ceramic.

Products containing both water and organic liquids (i.e., immiscible liquids) can be dried, and each volatile component will evaporate as if the other component were not present in the liquid mixture. However, in a ceramic, the organic component will interfere with the drying process in the initial phase of drying by blocking capillary movement of the water to the drying surface.

WARNING

Organic volatiles can result in release of flammable or explosive vapors in a dryer, resulting in fire and explosion.

When organic species, including common organic solvents, are removed in drying, flammable and explosive atmospheres can result. Operation below the *lower explosive limit* (LEL) for the solvent system is necessary to prevent ignition and/or explosion. Special ovens, called *Class A batch ovens,* are designed to accommodate unexpected ignitions without undue damage to property and injury to personnel. The use of organic solvents dictates the use of appropriate safety precautions in drying. The use of normal batch ovens with organic solvent systems is particularly dangerous because air is recirculated over heating elements at sufficient temperatures to cause unexpected ignition.

1.9 Dryer Stack Emissions and Air Pollution

Ceramic dryers have air flow originating at a hot supply (i.e., input of hot air) and terminating in an exhaust port or stack. Dryer stack emissions will contain water vapor and any other substances volatilized during drying. It is also possible that waste heat dryers will contain gases entrained in the hot supply gases, which in the case of waste-heat-fed dryers, could possibly contain kiln exhaust gases if any portion of the gases contain combustion products (as in the case of backdrafting of tunnel kilns).

The phenomenon of *scumming* (deposit of soluble materials on a ceramic product's surface after drying) is related to acid gas presence in dryers, including sulfuric acid (H_2SO_4). The consequence of scumming is discoloration or deposition of solubilized salts on the surface of the fired ceramic. However, scumming serves to illustrate that kiln gases, including SO_x, may be present in the dryer stack.

Other components of the dryer stack gases may be from volatilization within the ceramic. These include organic species used as binders and any products of reactions in the dryer environment.

1.10 Costs Involved in Drying Ceramics

Drying represents a process cost to the manufacturer in terms of handling and in terms of energy costs to accomplish drying. In direct-fired batch dryers, it is obvious that the energy to heat the hot supply air represents a real cost. In waste-heat-fed dryers, however, electrical energy costs can be appreciable. Recent applications of traveling propeller fans in brick drying (replacing conventional squirrel cage fans) have been one means of reducing electrical energy costs.

Ceramic dryers usually operate under positive pressure, meaning that the hot supply forced air creates positive pressure in the dryer. In cases where air

DRYING COSTS INCLUDE:

- Energy costs (thermal energy)
- Electrical costs to drive fans
- Capital costs (depreciation)
- Labor costs
- Maintenance costs
- Management or control costs
- Environmental costs

pollution control equipment is added to a dryer exhaust, part of the dryer may operate under negative pressure due to the use of induced draft fans. In these cases, air inspiration in the dryer can change local humidity, resulting in dryer problems.

1.11 Drying Defects

The drying process is the major source of defects in most ceramic products. These defects range from visible defects (cracks) to a reduction in physical properties in fired products in areas such as strength or elastic modulus. Defects also include those related to color and appearance, such as discoloration due to scumming.

The consequence of water removal in drying is ***shrinkage.***

The fundamental cause of most drying defects (cracks) is excessive drying shrinkage. Shrinkage typically ranges up to about 2–4% linear shrinkage for most plastic formed ceramic products. The goal in ceramic production is always to minimize drying shrinkage so as to minimize drying defects. The use of *grog* (non-shrinking aggregate) is one well-known way to limit drying shrinkage. However, the rate of drying is also related to shrinkage, with rapid drying favoring minimized shrinkage in many ceramics. Extended discussion on drying shrinkage is found in Chapter 3.

In many cases, drying tends to exaggerate defects created in ceramic products during the forming process (Chapter 7). Phenomena such as particle alignment, laminations, and residual forming stress may result in cracking upon drying. If any of these defects are created due to high forming rates or simply by convenience in forming, they result in cracking or warpage confirming one of Gil Robinson's favorite maxims:

What helps you in one place will hurt you in another.

PART B: ADVANCED CONCEPTS

1.12 Thermodynamic Considerations

The *third law of thermodynamics* can be used to define enthalpy:

$$dG = dh - TdS \tag{1.8}$$

where: dG = the incremental change in Gibbs free energy involved in a process.
dh = the incremental enthalpy change in a process.
T = the absolute temperature.
dS = the incremental entropy or nonrecoverable energy change in the process.

The third law can be restated as a word equation:

Total energy change = Exchangeable energy change + Nonrecoverable energy change

or,

Free energy change = Enthalpy change - Entropy change

Equation (1.8) can be rearranged to be:

$$dh = dG + TdS \tag{1.9}$$

where the enthalpy change imposed on a process goes to changing the total energy of the process plus the energy expended to create the process (the latter of which cannot be recovered). This means a real process cannot be 100% efficient (no possibility of recovery of 100% of energy expended) since some energy must be expended in a nonrecoverable form.

Entropy as nonrecoverable energy is usually explained in basic thermodynamics, such as energy expended in the form of frictional losses. In drying ceramics, one could visualize entropy expenditures in terms of slight particle translations or rotations during shrinkage of the ceramic, leading to frictional loss.

Enthalpy changes at constant pressure *(P)* are determined using the specific heat (Table 1.3) by integrating the relationship (summing all incremental changes):

$$dh = mC_p dT \tag{1.10}$$

where m is the mass (weight) of the substance being heated or cooled and C_p is the heat capacity of the substance (or system) at constant pressure.

Enthalpy changes at constant temperature *(T)* are determined using the relationship:

$$dh = pdV \tag{1.11}$$

where V is the volume of the material and dV is the incremental change in volume with a change in pressure.

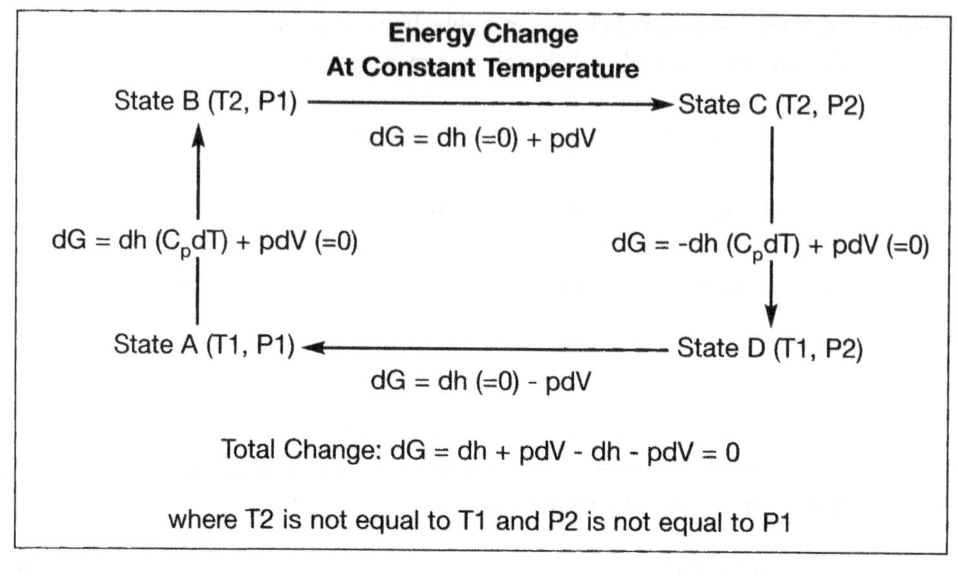

Figure 1.4: The Carnot cycle for an adiabatic process with changes in temperature (T) and pressure (p).

1.13 The Carnot Cycle

Energy changes in ideal physical processes may be explained using a Carnot cycle, illustrated in Fig. 1.4. Here the process is *adiabatic;* that is, there is no energy lost to or absorbed from the system of interest into or from the surroundings. The process is conducted with a system (perhaps an isolated volume of gas) from an initial (or reference) State A to a higher-energy State B through a temperature change at constant pressure. Then, the system is taken to a State C through a pressure change at constant temperature, then to a State D through a temperature change at constant pressure, and finally back to State A through a pressure change at constant temperature.

The key point is that the total energy involved in the process is constant in transition from State A through States B, C, and D, and back to State A, and the total energy change is zero. The Carnot cycle illustrates an ideal machine existing in an adiabatic world where there are no losses or gains of energy to the system.

While the ideal world of the Carnot engine does not exist, it is useful to use adiabatic conditions to illustrate drying situations. For example, we can add energy to an air segment, assumed under adiabatic conditions, and then we can later recover that energy, that is, use it in drying. With respect to the energy exchange in drying, assumption of adiabatic conditions on a limited scale (use of an isolated system) is convenient — even though the dryer as a large device is not an ideal machine and it does not exist in adiabatic conditions.

The Carnot cycle can be used to illustrate the fact that, in a total process

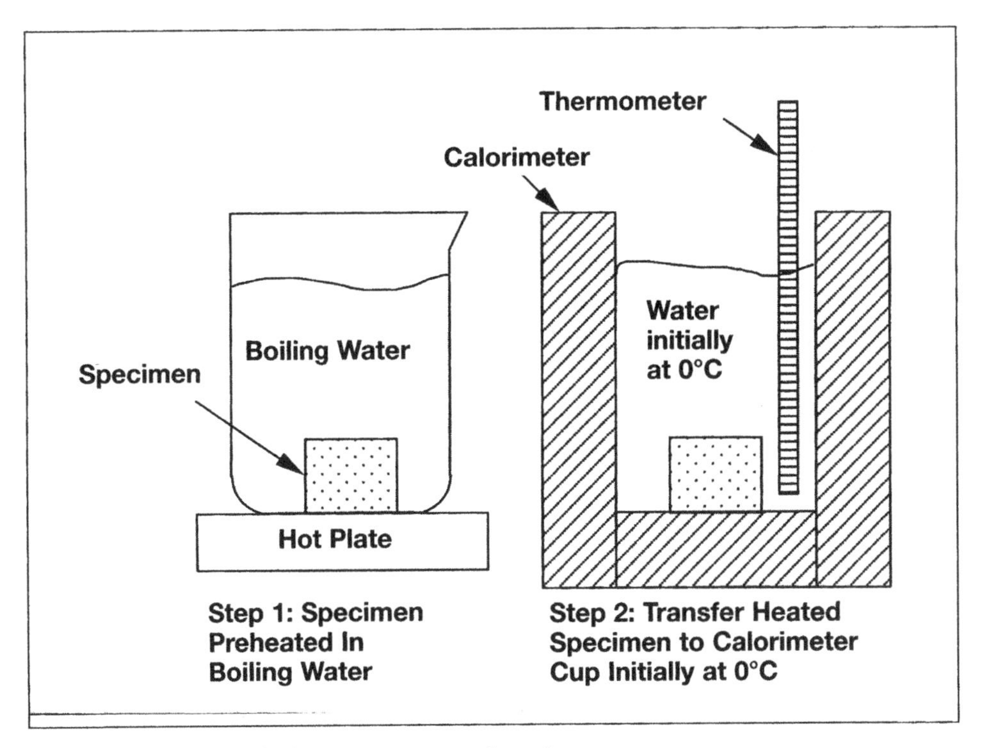

Figure 1.5: The calorimetry cup experiment.

involving evaporation beginning at a lower temperature, processing to a higher temperature, and returning to the initial temperature, the energy change is zero. This ensures that the temperature at which evaporation took place did not affect the total energy input or recovered. Another way of stating this with respect to adiabatic systems and evaporation is: "It doesn't matter how you got there — it only matters where you end up."

1.14 Thermodynamic Equilibrium

A key definition of equilibrium from thermodynamics is that it occurs only when $dG = 0$, that is, when no further changes are possible in the system in a real time frame. Thermal equilibrium occurs when no more heat transfer is possible, meaning no thermal gradients exist within the system of interest. Heat transfer in a near-adiabatic system is illustrated by the calorimetry cup experiment that is commonly used to determine the identity of a material by determining its specific heat (Fig. 1.5).

In its simplest form, the experiment consists of preheating an insoluble specimen using boiling water. Then, the specimen is quickly transferred to the calorimeter cup at an initially known temperature. To analyze the calorimetry cup experiment, it is necessary to realize that the final temperature of the system (cup + water + specimen) is to be determined by the investigator.

A simple heat balance equation is formulated as follows:
At equilibrium:

$$\Delta G = 0 = \Delta h - T\Delta S \qquad \textbf{(1.12)}$$

and $\Delta S \sim 0$ for such a simple experiment.

Therefore, after transfer of the preheated specimen to the calorimeter:

$$\begin{aligned}\Delta h &= \text{Heat gained by the water and the cup} \\ &\quad + \text{Negative heat gain (loss) by the specimen} \\ &= 0\end{aligned}$$

or

Heat lost by the specimen = Heat gained by the water and gained by the cup

If T_{final} is the temperature of the system at thermal equilibrium and m_{s} is the mass of the sample, m_{w} is the mass of water in the cup, and m_{c} is the mass of the cup, then:

$$(m_{\text{s}})(C_{\text{p,s}})(100°\text{C} - T_{\text{final}}) = (m_{\text{w}})(C_{\text{p,w}})(T_{\text{final}} - 0°\text{C}) + (m_{\text{c}})(C_{\text{p,c}})(T_{\text{final}} - 0°\text{C}) \qquad \textbf{(1.13)}$$

where the subscripts "c" and "s" represent the cup and the specimen, respectively.

The calorimeter cup was assumed to be in a well-insulated container; that is, there were no heat losses to the system or no heat gains from the system. The experiment is adiabatic for all practical purposes. At the final temperature for the system, T_{final}, the system is at thermal equilibrium, meaning that no further temperature changes are expected in the cup. In practical experiments, the student will observe the cup to reach a T_{final} that begins to change toward room temperature over time due to heat transfer to or from the room. Nevertheless, the point is made that the formulation of heat balance equations becomes a powerful tool in analyzing thermal processes — including dryer performance.

EXAMPLE 1.4: A calorimetry cup experiment is conducted with an unknown specimen weighing 25 g and employing an iron cup weighing 60 g filled with 100 mL of water. The specimen is held at 100°C (in boiling water) and is quickly transferred to the water filled calorimeter cup that is initially at 0°C. Assuming no boiling water was transferred to the cup along with the sample, the final temperature of the cup is determined to be 15.8°C. Using Table 1.3, what is the identity of the unknown specimen?

Solution:

The solution is found by inserting appropriate values into Eq. 1.13:

heat lost by the sample = heat gained by the water
+ heat gained by the cup

where $C_{p,s}$ (the heat capacity of the specimen) is to be determined.

$$(25\text{ g})(C_{p,s})(100°C) = (100\text{ g})(1\text{ cal/g})(15.8°C - 0°C) + (60\text{ g})(0.11\text{ cal/g})(15.8°C - 0°C)$$

and

$$C_{p,s} = 0.80$$

The sample is alumina according to Table 1.3.

Analyses of heat exchange in an adiabatic situation are very convenient to use in drying problems. The ability of air to cause drying in a ceramic can be illustrated in a simple problem where the heat lost by the air equals the heat gained by the ceramic. To make the problem manageable, the drying system is usually considered as adiabatic.

EXAMPLE 1.5: Given 4.5 kg of dry air at 90°C in a well-insulated container. Insert a green clay ceramic weighing 100 g at a moisture content of 10% wet basis at an initial temperature of 20°C. What is the final temperature in the oven, and what is the composition of the atmosphere?

Solution:

The ceramic consists of 90 g of dry clay and 10 g of water.

heat lost by the air = heat gained by the ceramic
+ heat used in evaporation
+ heat gained by water vapor

Let T_f = final temperature and c = clay, w = water, wv = water vapor, and a = air. The enthalpy of vaporization of water is denoted by Δh_v. Assume, for convenience, that all vaporization takes place at 20°C. Using data in Table 1.3:

$$(m_a)(C_{p,a})(90°C - T_f) = (m_c)(C_{p,c})(T_f - 20°C) + (m_w)(\Delta h_v) + (m_{wv})(C_{p,wv})(T_f - 20°C)$$

$$(4500 \text{ g})(0.24 \text{ cal/g°C})(90°C - T_f) = (90 \text{ g})(0.22 \text{ cal/g°C})(T_f - 20°C) + (10 \text{ g})(555 \text{ cal/g}) + (10 \text{ g})(0.46 \text{ cal/g})(T_f - 20°C)$$

So:

$$T_f = 84.3°C$$

The final composition of the air is 10 g of water in 4.5 kg of dry air since all water has evaporated.

Note that the sensible enthalpy loss by the air caused the air temperature to decrease. The energy lost by the air was completely transferred to the ceramic and the water it contained (it is adiabatic). This energy transferred to the ceramic caused evaporation (i.e., drying) to occur.

1.15 The Clausius-Clapeyron Equation

It can be shown that the work *(w)* involved in one circuit of the Carnot cycle (Fig. 1.4) is given by:

$$w = q\frac{T_2 - T_1}{T_2} \quad \textbf{(1.14)}$$

where q is the heat change involved in the cycle and the temperatures are those shown in Fig. 1.4. If the temperature difference $(T_2 - T_1)$ is small, the difference can be represented by a differential quantity dT. If changes in pressure are small, then $(p_2 - p_1)$ in Fig. 1.4 can be represented by the differential quantity dp.

The heat change at constant pressure for isobaric transitions in Fig. 1.4 is dh or Δh, and the energy change at constant temperature is $(V_2 - V_1)$. The area within the transitions shown in Fig. 1.4 is the work involved in completing the cycle or $(V_2 - V_1)$ dP (the area of the parallelogram).

Inserting these values in Eq. 1.14:

$$(V_2 - V_1)dP = \Delta H\frac{dT}{T} \quad \textbf{(1.15)}$$

which can be arranged to yield:

$$\frac{dP}{dT} = \frac{h}{T(V_2 - V_1)} \tag{1.16}$$

or simply:

$$\frac{dP}{dT} = \frac{\Delta h}{T\Delta V} \tag{1.17}$$

Equation 1.17 is called the Clausius-Clapeyron equation, and it is applicable to all types of transitions or changes in state of a material. The quantity Δh is the enthalpy of the process at the transition, T is the temperature of the transition, and ΔV is the volume change at the transition.

The Clausius-Clapeyron equation can be used to describe phase changes, including vaporization processes; this makes it useful in dealing with the evaporation of water in drying. For example, if the pressure involved is the vapor pressure of water, then Eq. 1.17 can be used to calculate the change in vapor pressure with changes in temperature. This is fundamentally useful in drying situations since we are usually encouraging evaporation by supplying heat. The Clausius-Clapeyron equation will be considered in subsequent chapters.

1.16 Problems

1. Given the following batch composition, express the composition on a dry and a wet weight basis:

Potters flint, –200 mesh	300 lb
Kaolin clay	400 lb
Feldspar	250 lb
Ball clay	50 lb
Water	22 gal

2. In a quick production-related test in a brick plant, you need to determine the moisture content of clay coming into the plant. You find a dry ceramic dish weighing 177 g to serve to hold a raw material sample. You put the dish on a platform balance and tare out the balance to read zero. Then you add raw material to the dish until it holds a net weight of 50 g of the clay.

   You place the dish with material in a microwave oven to obtain a quick dryout setting the microwave to run for 5 min. While you are temporarily out of the lab a coworker uses the balance, so you lose your automatic tare weight. You come back and continue drying the sample until a constant total weight for the dish and the dried raw material is 222 g. The dish is subsequently cleaned and found to weigh 175 g.

(a) What is the moisture content of the as-received clay on a wet and a dry basis?

(b) Name at least one factor that could lead to a likely error in your moisture determination.

3. Using geometrical calculations:

(a) Calculate the moisture adsorption on 1 μm diameter spherical particles of silicon dioxide assuming that a monolayer of water molecules covers the surface of the particles and that the "diameter" of the water molecule is the same as that of an oxygen atom. Express your answer in grams of water per gram of particles and as percent moisture on a dry weight basis.

(b) Repeat the calculation in (a) above for fumed silica with an average particle size of 0.01 μm assuming a spherical particle shape.

4. Express mathematically how the moisture adsorption of carbon fibers increases as the length *(l)* to diameter *(d)* ratio increases in the range $l / d = 1$ to $l / d = 20$.

5. Given the rheometer curve for a plastic clay shown below:

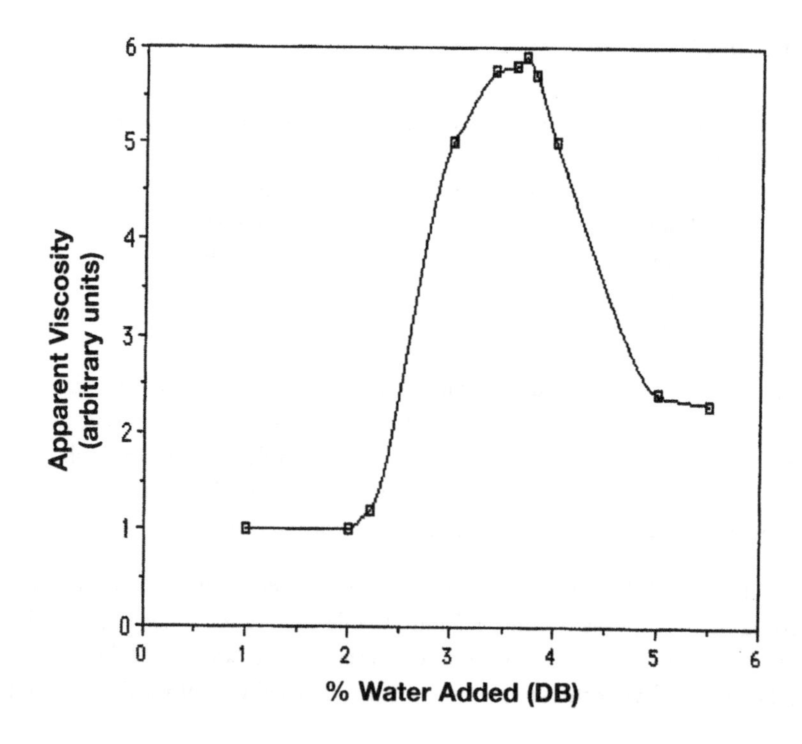

(a) On drying this raw material from a wet state (>5% moisture, DB), estimate the approximate moisture content (DB) where you expect the clay to stop shrinking.

(b) On continued drying, at what approximate moisture content do you expect the clay to have lost all of its pore and capillary water?

6. Calculate the specific heat of a composite ceramic with a phase analysis of 55% aluminum oxide, 40% aluminum, and 5% porosity.

7. Develop a relationship to show how the specific heat changes with increasing percentage of total porosity of a ceramic object, expressed in terms of bulk and true densities. The given information in the problem is:

Symbol/value	Definition
ρ	The bulk density of the ceramic object
ρ_t	The true or ultimate density of the ceramic object
C_p	The heat capacity of the porous ceramic (the unknown to be developed in this problem)
$C_{p,\,true}$	The heat capacity of the fully dense ceramic, i.e. the heat capacity at 0% porosity
0.24 cal/g°C	The heat capacity of dry air

8. A calorimeter cup experiment is conducted to determine the room temperature specific heat of a ceramic. Details of the experiment are as follows:

Weight of the ceramic	275 g
Weight of the calorimeter (cup)	100.2 g
Calorimeter material	Aluminum (specific heat 0.217 cal/g°C)
Volume of water in the calorimeter cup	150 mL
Initial temperature of cup and water	20°C
Initial temperature of specimen	100°C
Final temperature of the cup	42°C

(a) What is the specific heat of the ceramic?

(b) How much enthalpy did the specimen lose after being transferred to the cup?

(c) How much enthalpy did the calorimeter and water gain after the specimen was transferred to the cup?

9. Five pounds of green brick produced at 8% moisture content WB at a temperature of 70°F are placed in a large styrofoam container containing 30 lb of dry air initially at 200°F.

(a) In a practical time frame of several hours, what is the final temperature of the air when thermal equilibrium is attained within the container assuming good air circulation and no heat losses from the container?

(b) How is sensible heat lost from the air?

(c) How much latent heat does the air gain?

10. In an undergraduate lab, you build a small furnace using 10 lb of high-alumina castable refractory based on 80% tabular alumina and 20% high-purity calcium aluminate cement by weight. You anticipate operating the furnace at 1600°C in the small interior chamber, and a similar furnace in your lab is found to exhibit a cold face (exterior) temperature of 300°C. Given the heat capacity of alumina and calcium aluminate (as $CaO{\cdot}Al_2O_3$) as follows:

General relationship:

$$C_p \text{ (J/mol K)} = a + b\,(10^{-3}T) + c\,(10^{-3}T^2) + d\,/\,(10^{-3}T)$$

where:

Factor	Al_2O_3	$CaO{\cdot}Al_2O_3$
Formula weight	101.96	158.04
a	+154.96	+236.00
b	−16.168	−39.240
c	+7.120	+14.360
d	−20.817	−32.120

If the average refractory temperature is 950°C, how much heat is required in operating the furnace just to heat the furnace itself to the operating temperature?

11. If an organic liquid immiscible with water is used along with water in the form of an emulsion in fabricating a ceramic, explain the effect in drying with respect to the following:

(a) Vapor pressure of water within the ceramic during drying. (Hint: Consult a physical chemistry textbook on vapor pressure and distillation of immiscible liquids.)

(b) Rate of drying.

12. A plant is producing 12 × 12 in. Spanish tile by extruding a red firing shale body using stiff extrusion equipment. The tiles are set on kiln cars laid flat in piers (like a deck of cards laying on a table) to a setting height of 15 pieces. This setting pattern is used to allow for flashing or reduction firing to create a characteristic periphery of color on the exposed face of the tile.

However, in drying there is excessive loss due to the formation of vertical cracks in the center of the exposed edge of the tile in the setting. Without using a new procedure to dry the tile that might result in double handling of the material (resetting the tile for firing), discuss methods to reduce the cracking problem.

13. In manufacturing ceramic products using tunnel kilns, waste heat from cooling is normally used in drying the product. It is usually desired to avoid allowing combustion gases from entering the dryers due to their potential content of acid gases such as H_2SO_4 or HF. If these enter the dryer, metal salts can be created in the drying product, which migrate to the surface and lead to a discoloration on the fired product known as scumming. Discuss ways to determine if acid gases are entering the dryer.

14. Given Example 1.5:

(a) How much sensible enthalpy is lost by the air in drying?

(b) How much latent enthalpy is gained by the air in drying?

CHAPTER 2

Water, Air, and Water Vapor

PART A: WATER

2.1 Introduction

Water is the familiar liquid phase of the compound dihydrogen oxide (H_2O). It is a well-known essential for life on earth and it plays a key role in fabrication processes for ceramics. Ceramic raw materials, as well as rocks, may contain chemically combined "water" (actually hydroxyl units) that is released as steam during heating of the ceramic.

Water is a liquid phase at ordinary temperatures, and it provides for the plasticity exhibited by many particulate masses when wet. This plasticity permits forming operations like molding and extrusion to be possible. In turn, water must be removed before heat treating or firing of the ceramic. If water is not removed prior to firing, cracking is likely and steam spallation (steam explosions) may occur. An understanding of the fundamentals of water chemistry and physics aids in controlling both the forming and drying operations in ceramic processing.

In introductory chemistry courses, the water molecule is described as consisting of an oxygen atom coordinated with two hydrogen atoms spaced apart at a bond angle of 105°. This angle is attained due to the orbital relationships of the oxygen and hydrogen atoms. The atoms are linked with strong covalent bonds, making water a very stable substance.

The free energy of formation of water is compared to ceramic compounds in Table 2.1, which shows that water is a relatively stable substance. Note that a more negative free energy of formation implies greater stability of the compound. This is also common sense, since the only usual process in which the average citizen sees water broken into its constituents is during electrolysis (as is possible in charging a battery).

Water is both essential to life and essential in forming most ceramic products. Water activates plasticity in most particulate systems. Water must be removed before firing or sintering.

Table 2.1: Free energy of formation (ΔG_0) of water and oxides at 25°C (kcal/mol)

Copper oxide (Cu_2O)	–38.1
Lead oxide (PbO)	–43.9
Water (H_2O)	–54.6
Iron oxide (FeO)	–59.4
Iron oxide (Fe_2O_3)	–179.1
Silicon dioxide (SiO_2)	–190.4
Aluminum oxide (Al_2O_3)	–376.8

The water molecule is small, having a diameter of approximately that of the oxygen atom of 1.4 Å (1.4×10^{-10} m). The molar volume of water at 20°C is approximately 18 cm^3 or 18 mL. This implies that there are an enormous number of water molecules in a typical sip of water from a glass (6×10^{23} or about one-half million-billion-billion in 18 mL [about 0.5 U.S. fluid ounces]).

A very important feature of the water molecule is that it exhibits a dipole moment. In other words, water molecules exhibit a positive side and a negative side (Figure 2.1). The polar nature of the water molecule means that it has a high dielectric constant, because molecules will align in an applied electric field. Water is a good solvent for most ionic substances, because the water can shield charged species from each other, allowing them to stay in solution. The alignment of water molecules with

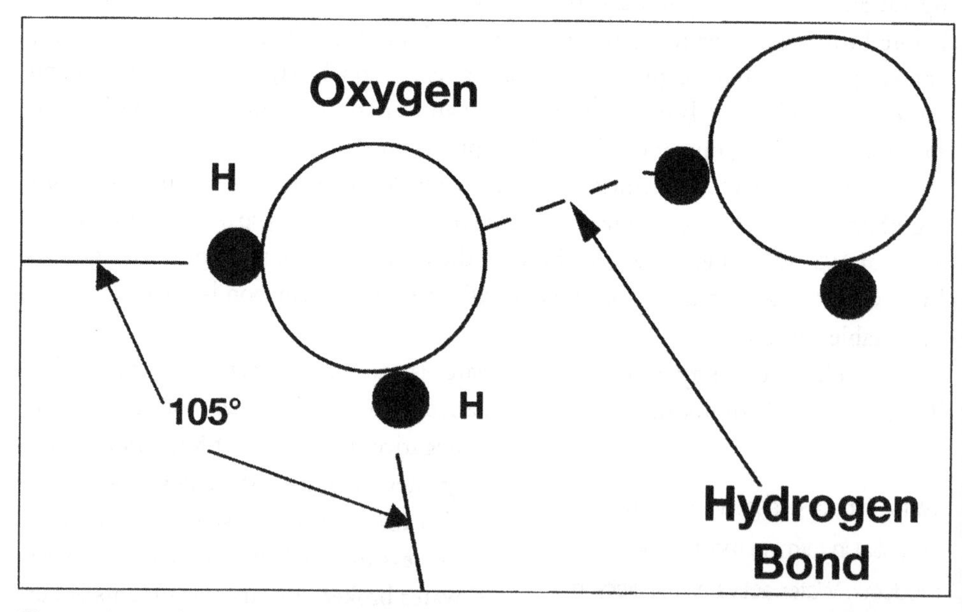

Figure 2.1: Conventional representation of water molecules showing polar nature and hydrogen bonding.

applied electric fields is the basis for dielectric drying techniques (Chapter 8).

The polar nature of the water molecule is also why the positive side of the water molecule is attracted to most oxide surfaces. These surfaces usually appear from an outside vantage point as negative because of the electronic charge cloud of surface oxygen atoms. Water does not want to completely spread out or wet an oxide surface, as there is also repulsion between the negative side of the water molecule and the surface. Some oxide surfaces, such as the edge of kaolin crystals, also exhibit positive surface charges that attract the negative side of the water molecule and repel the positive side of the water molecule.

Many of the properties of water are a result of *hydrogen bonding*, a type of chemical bonding formed between the electronegative oxygen atom in one water molecule and hydrogen atoms in an adjacent water molecule. These hydrogen bonds are much weaker (energy of formation ~10 kcal/mol) than the covalent O–H bonds in the water molecule (energy of formation –110 kcal/mol), and hydrogen bonds are constantly being broken and remade within the liquid as the molecules move around. From this standpoint, a viewpoint of water as a rigid assemblage of molecules is incorrect, and making an analogy between water molecules in motion and ants in a disturbed nest might be more realistic.

EFFECT OF HYDROGEN BONDING IN WATER:

- Boiling point at 100°C.
- High surface tension.
- Attraction to surfaces.
- Density

Hydrogen bonding influences many of the properties of water near room temperature and that also affect ceramic processing. One of the most important is that water would be expected to boil (change state from a liquid to a vapor) at about –80°C instead of 100°C at 1 atm pressure if hydrogen bonding did not occur.

The boiling point of water is thus a consequence of hydrogen bonding, which tends to keep the molecules in the liquid state until sufficient energy is attained by the molecules to escape into the vapor phase. Hydrogen bonding also influences the surface tension or surface energy that must be overcome for vaporization (evaporation) to take place.

2.2 Density

The density of water reaches a maximum at 4°C (Table 2.2), and the density of ice is 0.92 g/cm^3, explaining why ice floats on a frozen pond. The slight expansion of water as it cools below 4°C is due to the alignment of molecules just before the water changes state to become crystalline. At 4°C, 1 g of water occupies a volume of 1 cm^3. It is common for engineers to use 1 g/cm^3 as the density of water at room temperature, but this is actually in error by a slight amount (~0.2%), as shown in Table 2.2.

Table 2.2: Density of water at 1 atm pressure

Temperature (°C)	Density (g/mL)
0	0.99987
4	1.00000
20	0.99823
40	0.99224
60	0.98324
80	0.97183
100	0.95838

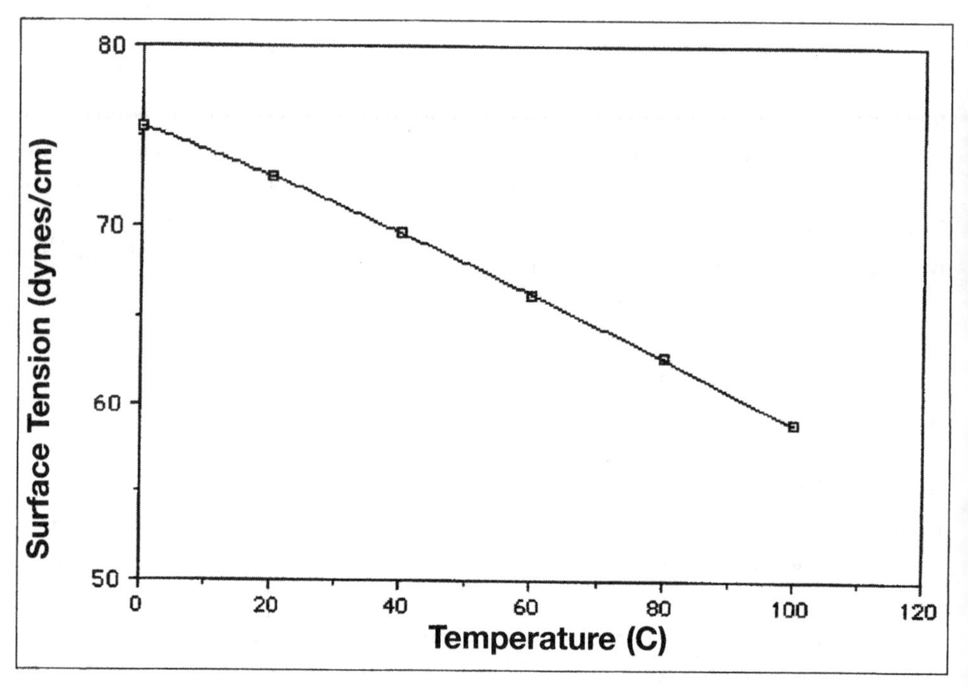

Figure 2.2: Surface tension of water.

2.3 Surface Tension

The surface tension of water is illustrated in Fig. 2.2, and it is given in Table 2.3 along with values for other liquids. Surface tension is the force acting to maintain a constant or level surface in a container. Its effect is to create water droplets when water is poured or sprayed into air since a sphere has the minimum surface area (minimizing surface forces). The uniform decrease in surface tension of water with increasing temperature is partly due to the ease with which hydrogen bonds can be broken as temperature increases, and it implies that water has an easier time of evaporation as temperature increases — as is observed.

Table 2.3: Surface tension of various liquids against air (dyne/cm)

Substance	0°C	20°C	40°C	60°C	80°C	100°C
Water	75.6	72.8	69.6	66.2	62.6	58.9
Acetone	26.2	23.7	21.2	18.6	16.2	
Ethanol	24.0	22.3	20.6	19.0		
Benzene	31.6	28.9	26.3	23.7	21.3	

2.4 Surface Tension and Capillary Suction

The action of water rising in a capillary in apparent defiance of gravity is illustrated in Fig. 2.3. The process in the capillary tube is wetting the inner surface of the tube, with hydrogen bonds playing a role in attracting the water molecules to the surface of the tube. Since water is attracted to the surface, it continues to rise on the walls, creating a suction or force *F* that causes the tube to fill. The water is seen to rise in the capillary to reach an elevation *h*, forming a meniscus (water level) with a contact angle created as the water tries to continue wetting the capillary surface. The forces may be resolved in the vertical or *y* direction using trigonometry.

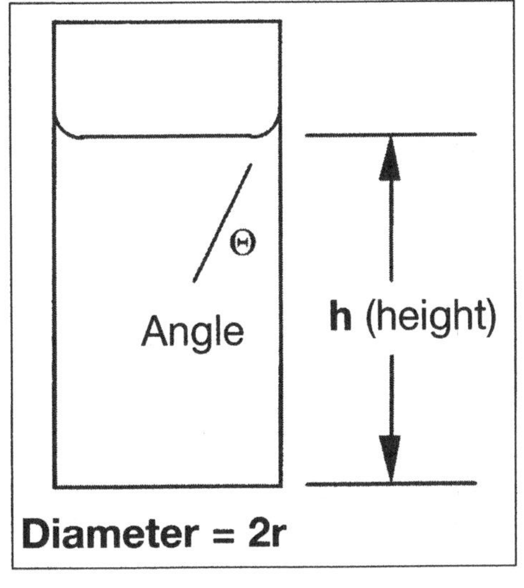

Figure 2.3: Rise of water in a capillary.

The suction (upward) force is:

$$F_y = 2\pi r \gamma \cos\Theta \quad \textbf{(2.1)}$$

where: r = radius of the capillary.
$2\pi r$ = circumference of the capillary.
γ = surface tension of the water against the capillary.
Θ = contact angle (Fig. 2.2).

The downward force due to the acceleration of gravity is:

$$F_y = \pi r^2 h \rho g \quad \textbf{(2.2)}$$

where: h is the height of the water in the capillary.
ρ is the density of the water.
g is the gravitational constant.

At equilibrium, the upward and downward forces balance, so:

$$2\pi r\gamma \cos\Theta = \pi r^2 h\rho g \tag{2.3}$$

and:

$$h = \frac{2\gamma \cos\Theta}{r\rho g} \tag{2.4}$$

If we use diameter of the capillary, *d,* in place of the radius and recognize that as Θ approaches 0°, $\cos\Theta \sim 1.0$, then:

$$h = \frac{4\gamma}{d\rho g} \tag{2.5}$$

It is very easy to see that as the diameter of the capillary decreases, the height of rise in the capillary increases. In porous ceramic products, water will be absorbed due to capillary suction resulting from action of pores generally smaller in diameter than about 10 μm.

As will be seen in Example 2.1, pores with a diameter less than 1 μm particularly affect the rise of water within a capillary. The property of ceramic products to absorb water from a reservoir is called *capillarity,* and in common language, the product is said to exhibit *suction* by drawing water into the product itself. Practical examples of suction include the withdrawal of water from a mortar bed when brick are laid by a mason, and brick in a wall becoming wet during periods of rain. In another example, refractory concrete will lose water by capillary suction if cast against old or existing porous concrete. In this case, plastic film should be used to isolate the newly poured concrete from the old concrete so that proper hydration in the new concrete takes place.

EXAMPLE 2.1: In a system at 20°C at sea level, calculate the rise in height of water in a capillary of the following diameters:

(a) 1 mm or 10^{-3} m

(b) 100 μm or 10^{-2} cm

(c) 1 μm or 10^{-4} cm

Note: 1 μm = 10^{-6} m = 10^{-4} cm = 10^{-3} mm

Given ρ = 0.99823 g/cm³, γ= 72.8 dyne/cm or 0.0728 g/s², and

$g = 981$ cm/s^2, then

(a) For a 1 mm capillary:

$$h = \frac{4\gamma}{d\rho g} = \frac{4\left(0.0728\ \mathrm{g/s^2}\right)}{\left(10^{-1}\ \mathrm{cm}\right)\left(0.99823\ \mathrm{g/cm^3}\right)\left(981\ \mathrm{cm/s^2}\right)} = 0.00297\ \mathrm{cm}$$

(b) For a 100 μm capillary, using the same formula: h = 0.0297 cm (~0.0 in.)

(c) For a 1 μm capillary, using the same formula: h = 2.974 cm (~1.2 in.)

2.5 Viscosity

Viscosity is a key property of a fluid and is defined in the simplest terms as the resistance of the fluid to shear (in terms of force per unit area in one volume segment relative to another where the relative velocities in both fluid segments differ by unity). It is often commonly referred to as the internal friction or cohesion of the fluid. It follows by the Poiseuille equation that the property of the apparent viscosity η is related to the volumetric flow v per second through a tube of radius r and length L under a pressure p as follows:

$$v = \eta \frac{8L}{\pi r^4 p} \tag{2.6a}$$

Rearranging Eq. 2.6, we obtain an expression for the pressure required to obtain a certain flow in a pipe:

$$p = \eta \frac{8L}{\pi r^4 v} \tag{2.6b}$$

where η is a proportionality constant in the Poiseuille equation. There are other ways to define viscosity. The use of the Poiseuille equation is a simple way of introducing the concept of viscosity.

If volume is in cm^3/s, L and r are in cm, and p is in dynes/cm^2, the dimension for η is the poise (dyne-s/cm^2). The most common unit used is the centipoise (1 cp = 0.01 p). The term *apparent viscosity* is used with Eqs. 2.6a and 2.6b because of the phenomenon of nonlaminar flow in pipes, and it is the property obtained from laboratory measurements (where some degree of nonlaminar flow cannot be avoided).

Absolute viscosity is a property of perfect fluids that exhibit laminar flow; such

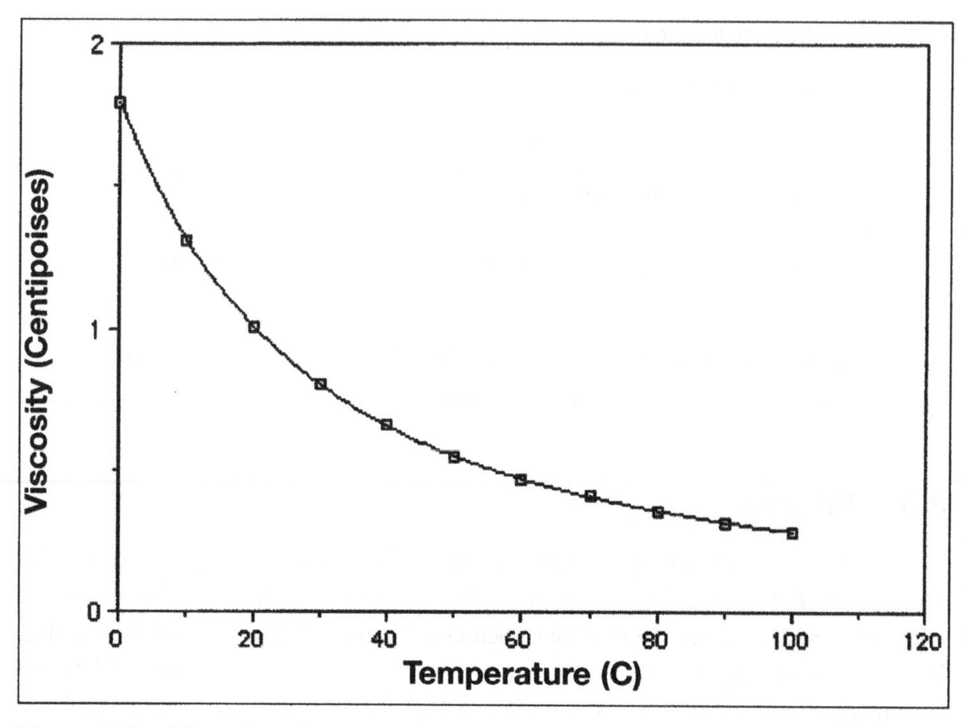

Figure 2.4: Viscosity of water.

Table 2.4. Viscosity of water as a function of temperature

Temperature (°C)	Viscosity (cp)
0	1.792
20	1.005
40	0.656
60	0.469
80	0.357
100	0.284

fluids are also called *Newtonian fluids*. *Kinematic viscosity* is defined as the absolute viscosity divided by the density of the fluid. The unit of kinematic viscosity is the *stoke* (1 cm^2/s).

The viscosity of water exhibits a remarkable decrease as temperature is increased (Fig. 2.4 and Table 2.4). The viscosity decrease from room temperature (20°C) to the boiling point is an incredible 71%. Stated another way, using Eq. 2.6, with all other factors equal, 100°C water will exhibit a flow rate through a tube some 3.5 times greater than 20°C water. Flow rates through small diameter tubes, i.e. capillaries, have an important affect on drying rates (Chapter 3).

Table 2.5: Saturation vapor pressure for water in the low-temperature region

Temperature (°C)	Vapor pressure (mm Hg)
0	4.579
20	17.535
40	55.324
60	149.38
80	355.1
100	760

2.6 Vapor Pressure

Evaporation is the process of changing the state of a liquid to a gas through the application of thermal energy. The gas produced by the evaporation of water is called water vapor. The application of thermal energy to water can cause some of the molecules to overcome their attraction for others and escape into the surrounding atmosphere. The reverse of the vaporization process is condensation. Condensation results when the kinetic motion of gaseous molecules is reduced through the extraction of energy, and the vapor molecules convert back to the liquid state.

The vapor generated by the evaporation of water exerts a pressure against the walls of any confining enclosure. This pressure increases with time from the commencement of evaporation up to a maximum. Then there is no further increase in pressure with time as long as the temperature or total pressure of the environment does not change. This final pressure is called the *saturated vapor pressure of water* of that particular temperature. There can be no further net transfer of material to the vapor state after the saturated vapor pressure is attained.

Evaporation of water continues until the saturation vapor pressure is achieved.

An increase in temperature of the environment would result in a resumption of evaporation until a new maximum vapor pressure is reached. The saturated vapor pressure is a function of temperature (Table 2.5 and Fig. 2.5).

The vapor pressure of a mixture of gases is the sum of the pressure of each component gas. Each gas can exert its characteristic saturated vapor pressure, assuming a sufficient supply of each component and equilibrium conditions. This may be illustrated for air and water vapor by starting with an evacuated unrestrained enclosure at zero internal pressure. Introduction of water into the enclosure at 20°C, in sufficient quantity, will develop the saturated vapor pressure of water, which is 2.3 kPa (0.689 in. of mercury or 0.0249 psi) as the total pressure within the enclosure. Adding atmospheric air at a pressure of 1 atm to this enclosure will further increase the internal pressure by 101.3 kPa (29.92 in. of mercury) to a total value of 103.6 kPa.

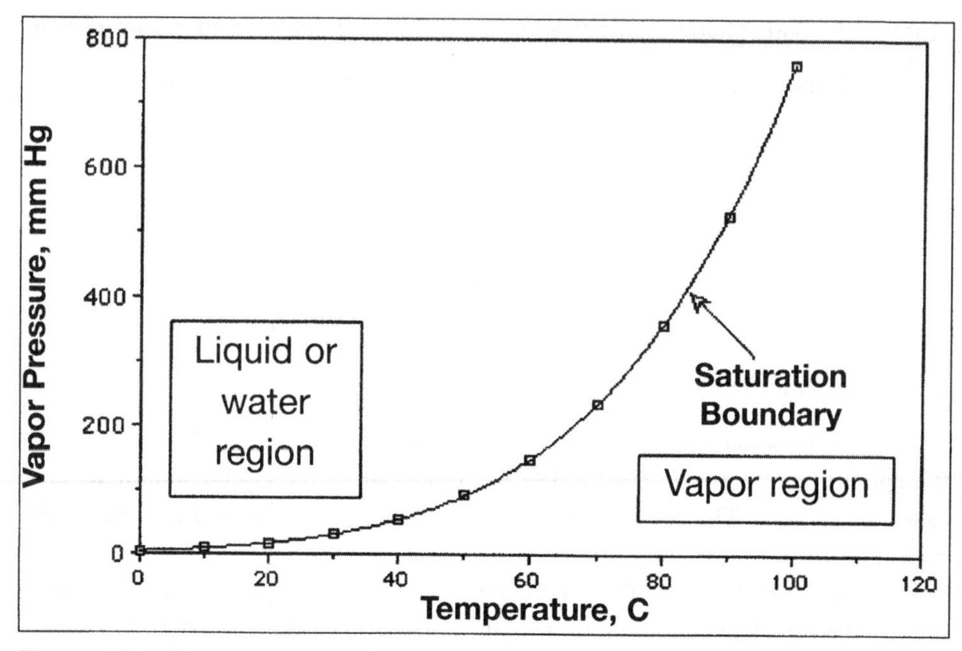

Figure 2.5: Vapor pressure of water in the low-temperature region.

The addition of another liquid has the potential of a further increase in total pressure by the amount of saturated vapor pressure of the third component. If, however, there is a restriction on total pressure in the enclosure, components within the mixture must be restricted in their evaporation. This is the fundamental reason that there are limits on the amount of water that air can carry or hold at any temperature.

The saturated vapor pressure changes between materials and is characteristic of a particular substance. The extraction of heat from an atmosphere saturated with vapor will produce a condensation of the vapor.

Using Eq. 1.17, it can be shown that the vapor pressure at saturation, P_{ws}, follows the relationship:

$$P_{ws} = P_{w,373K} \exp \frac{-\Delta h_v}{\frac{1}{T} - \frac{1}{373K}} \tag{2.7a}$$

where $P_{w,373K}$ is the vapor pressure of water at 100°C (373 K). The vapor pressure at saturation is also called the equilibrium vapor pressure. As shown in Fig. 2.5, the vapor pressure reaches a maximum of 1 atm at 100°C. The saturation boundary in Fig. 2.5 cannot be crossed in a vertical direction without condensation occurring.

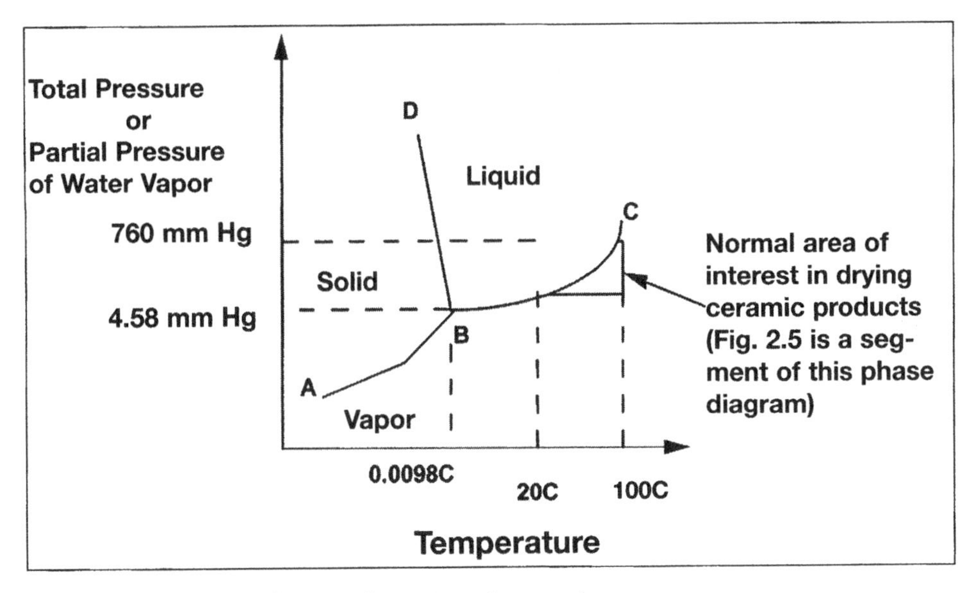

Figure 2.6. Phase diagram for water at normal pressures.

2.7 Phase Relationships for Water

Water can exist as a solid, liquid, or vapor at normal pressures when encountered in our environment. A phase diagram for water is shown in Fig. 2.6.

Several two-phase equilibria are illustrated as follows:

- Equilibrium between ice and water: Line B – D (the well-known process of fusion or melting of ice). Note a relative insensitivity of the freezing point to applied pressures.
- Equilibrium between water and water vapor: Line B – C. This is the saturation line, the pressure that, if exceeded, will result in condensation or the presence of liquid. *This is the exact same relationship as in Fig. 2.5, and the boxed area is the region of the phase diagram of normal interest in drying of ceramics.*
- Equilibrium between ice and water vapor: Line A – B. This is the sublimation line, the phase boundary for the vaporization of the solid.

Point B is the *triple point of water,* the only place where water, water vapor, and ice are at equilibrium. The coordinates of this point are 4.58 mm Hg pressure and 0.0098°C temperature, and, as such, the triple point is not of interest in ceramic drying processes.

The dotted line at 760 mm Hg illustrates heating ice at normal atmospheric

Table 2.6: Vapor pressure reduction over NaCl–water mixtures (addition of salt in mol/L where 1 mol of NaCl is 58 g)

Mol NaCl/L	Vapor pressure reduction (mm Hg)
0.5	12.3
1.0	25.2
2.0	52.1
3.0	80.0
4.0	111.0
6.0	176.5

pressure. Starting from the left, the equilibrium phase is ice, and on crossing the phase boundary, ice melts (0°C). Continuing to the right, liquid water is heated, evaporating during heating if the air is not saturated in water vapor, until the water boils (at 100°C). As the water vapor is heated (above 100°C), only that single phase of water vapor can exist (at 1 atm pressure). The saturation line (or phase boundary) takes on special significance in the discussion of psychrometry (Chapter 4).

2.8 Effect of Solutes (Dissolved Salts) on Vapor Pressure

The addition of a salt to a liquid lowers its vapor pressure in any particular temperature. The liquid molecules at the surface of the liquid are replaced in part by the dissolved salt species molecules. The result is a lower rate of evaporation per unit of surface area of the liquid because salt species (hydrated metallic ions) exert an attractive force on water molecules. Therefore, at equilibrium the rate of condensation is proportionately lower.

Addition of salts effectively results in a depression of the saturated vapor pressure at a particular temperature, and it also means that a higher temperature must be used to bring the solution to the boiling point. The extent in reduction of the vapor pressure is a function of the concentration of the dissolved substance.

The reduction in vapor pressure of water by the addition of sodium chloride is given in Table 2.6. Since the water vapor pressure is reduced in a closed container containing a salt such as NaCl, the relative humidity is also reduced over the salt solution. Salt solutions are conveniently used to maintain constant humidities in closed containers in the laboratory during drying experiments.

Vapor pressure also can be lowered by the attraction between water and other substances such as clay minerals. This attractive force holding the water molecules effectively reduces the vapor pressure and increases the temperature requirement for evaporating this water. For this reason, some clay minerals require temperatures exceeding 100°C for complete drying to be accomplished.

Further reduction in vapor pressure is obtained when water is confined to small capillaries. The greater the curvature of the water film at the walls of the capillary, the lower the vapor pressure. This can result in a very sizable reduction in vapor pressure in the capillaries of a drying object. Therefore, more energy is required to evaporate water from small pores than from large pores. Liquid movement during drying tends to go from big pores to small ones, which retards drying or increases the energy requirements for the final stages of drying.

The effect of dissolved salts in a clay–water system is complicated by the fact that dissolved salts will result in opposite effects on surface tension and on viscosity. Because the concentration of dissolved salts is usually small in wet ceramic masses (green or undried parts), the surface tension may be affected to a greater extent than viscosity, suggesting a slight depression in drying rate.

PART B: AIR

2.9 Composition of Air and Air Pollution

The average composition of dry air is given in Table 2.7, where it is seen that nitrogen and oxygen make up ~99% of air by volume at the earth's surface. Several facts indicate that air is a mixture rather than a compound:

- The composition of air is variable.
- If liquid air is boiled, nitrogen is the first principal constituent to be released.
- Air can be produced from a mixture of the normal constituents with no heat evolution or absorption as would be expected if air were a compound.

It is well known that oxygen is necessary for the maintenance of life and that the earth's crust contains a preponderance of oxygen (~49.2%). As a gas, oxygen is colorless, odorless, and tasteless. Its chemical importance is due mainly to its participation in oxidation reactions, that is, common combustion.

The free energies of formation of selected common oxides at room temperature are given in Table 2.8. Note that all given numbers are negative, indicating the energy is given off when the reaction takes place indicating the reaction is spontaneous or will readily proceed. These data explain why most metals are present in the earth's crust as oxide species.

Because air is a mixture, it can be

PRIMARY CONSTITUENTS OF AIR:

- 78% N_2 (0.78 atm partial pressure)
- 21% O_2 (0.21 atm partial pressure)

Table 2.7: Average composition of dry air (vol%), boiling points of selected constituents (K), National Ambient Air Quality Standards (NAAQS) in the United States, and indoor air standards or threshold limit value (TLV) in the United States

Species	Chemical formula	Voi%	Boiling point (K)	NAAQS (ppm)	TLV (ppm)
Nitrogen	N_2	78.04	77.3		
Oxygen	O_2	20.99	90.2		
Argon	Ar	0.94	87.4		
Carbon monoxide	CO	0.01		9[a]	50
Carbon dioxide	CO_2	0.023–0.050	Sublimes		5000
Ozone	O_3	0.02		0.08[b]	0.1
Hydrogen	H_2	0.01	20.4		
Neon	Ne	0.0015	27.2		
Sulfur dioxide	SO_2	0.002		0.03[c]	5
Nitrogen dioxide	NO_2	0.001		0.053[c]	5
Helium	He	0.0005	4.2		

[a] *1-h average.*

[b] *8-h average adopted 6/97 replacing old level of 0.12 (1-h average).*

[c] *Annual arithmetic mean.*

Note: Particulate matter is included in the NAAQS as a criteria pollutant.

Table 2.8: Selected free energies of formation (ΔG_f) of metal oxides at 25°C

Species	ΔG_f at 25°C (kcal/mol)
Al_2O_3	–376.8
CaO	–144.3
$CaO{\cdot}SiO_2$	–357.5
CO	–32.81
CO_2	–94.26
FeO	–59.38
Fe_2O_3	–179.1
MgO	–136.2
$MgO{\cdot}SiO_2$	–326.7
SiO_2	–190.4

mixed with other gases or particulate matter, such as those originating from smokestacks or other emission points in manufacturing. When the admixed gases cause a threat to human health or welfare, they are commonly known as air pollution. Air pollution can exist in many forms: odors, hazardous gaseous species, condensable particulates, and particulate matter (solid) forming an aerosol.

Ozone (O_3) is an allotropic form of oxygen formed by the action of an electrical discharge, the action of ultraviolet light on oxygen, or by photochemical processes. In the upper atmosphere (stratosphere), ozone plays the essential role of absorbing harmful radiation. Near ground level, ozone is formed by the interaction of light, nitric oxide (primarily from automotive exhaust), and hydrocarbons (plants and air pollution) in the photochemical process.

Under the Clean Air Act in the United States, air pollutants are divided into two categories:

- Criteria pollutants: pollutant species that have an adverse effect on public health or welfare. Examples of interest in ceramic manufacturing are species listed in the National Ambient Air Quality Standards such as CO, NO_2, O_3, Pb, SO_2, and particulate matter.
- Air toxics or hazardous air pollutants: pollutant species that cause an increase in mortality or an increase in serious illness. Species of interest in ceramic manufacturing include hydrogen fluoride (HF) and hydrogen chloride (HCl).

Suspended particulate matter with an aerodynamic diameter greater than about 30 μm can remain airborne only for short periods, whereas particles in the range 1-10 μm can remain suspended for 100–1000 h. Particulate matter with an aerodynamic diameter less than about 10 μm can enter the lungs, and current scientific thought is that finer particulate matter causes injury to the health of persons with respiratory difficulties (see, for example, *Federal Register,* Vol. 61, No. 241 [13 December 1996]). Scientific uncertainty remains over just how fine particles must be before physiological injuries increase dramatically, but the EPA has set 2.5 μm as a criterion for future regulation of fine particulate matter.

AIR POLLUTION CATEGORIES:

- Criteria pollutants
- Air toxics or hazardous air pollutants

2.10 Properties of Air

The pressure of the air in the atmosphere at sea level, referred to as *normal atmospheric pressure,* is given in Table 2.9. Pressure in ductwork, dryers, and kilns has been traditionally measured with an inverted U-tube manometer, which provides pres-

Table 2.9: Atmospheric pressure at sea level

1 atm or 1.0133 bar
101,325 N/m^2 or 0.101325 N/mm^2 (MPa)
10,333 kg/m^2
14.70 lb/in.^{2}
1,013,250 dyne/cm^2
760 mm or 76 cm Hg (height of mercury column) or 29.92 in. of Hg
33.9 ft of water (height of water column) or 406.8 in. of water

Table 2.10: Density of air at 1 atm and near room temperature

Temperature (°C)	Density (g/mL)	Density (lb/ft^3)
0	0.001298	0.0810
10	0.001247	0.0778
20	0.001205	0.0752
30	0.001165	0.0727

Table 2.11: Absolute enthalpy of dry air

Temperature (°C)	Temperature (°F)	Enthalpy (kJ/kg)	Enthalpy (Btu/lb)
7	45	280.1	120.4
27	81	300.2	129.1
77	171	350.5	150.7
127	261	401.0	172.4
227	441	503.1	216.3
327	621	607.0	261.0

sure in millimeters of mercury or inches of water column. Inches of water column (WC) is a convenient measure due to the relatively low pressures used in heated ceramic process equipment, and derived units are useful (1 in. WC = ~0.036 lb/in.^{2}).

The density (ρ) of dry air in g/mL is given by the relationship:

$$\rho = \frac{0.001293P}{76(1 + 0.00367T)} \quad \textbf{(2.7b)}$$

where the temperature *(T)* is in degrees Celsius and the pressure *(P)* is in cm of mercury. Selected data are shown in Table 2.10.

Enthalpy values for dry air are given in Table 2.11 for 1 atm pressure at selected temperatures of interest in drying. These values are calculated from enthalpies of air constituents given the assumption that air is an ideal gas — which may be reasonable at low temperatures and low pressures in drying.

EXAMPLE 2.2: Calculate the enthalpy required to heat 1 kg of dry air from 77 to 227°C, using the tabulated enthalpy data in Table 2.9. Compare this result to the enthalpy required based on the specific heat of air in Table 1.3 (assuming for this problem that the specific heat of dry air is constant over the temperature range of interest).

(a) Calculation based on absolute enthalpy:

Enthalpy at 227°C	503.1 kJ/kg or J/g
– Enthalpy at 77°C	– 350.5 kJ/kg
	152.6 kJ/kg = 152.2 J/g

(b) Calculation based on specific heat (C_p of dry air = 0.24 cal/g°C or 1.00 J/g°C from Table 1.3):

$$\Delta h = mC_p\,\Delta T$$

$$\Delta h = (1000\text{ g})\,(1.00\text{ J/g°C})\,(227 - 77\text{°C}) = 150\text{ kJ/kg}$$

The small difference in results is primarily due to the assumption of constant specific heat for air over the temperature range of the problem in calculation (b).

PART C: AIR–WATER VAPOR MIXTURES

2.11 Introduction

The study and analysis of air-water vapor mixtures is called *psychrometry.* In a like manner, the vapor pressure relationships in drying are illustrated on charts, such as those in Figs. 2.5 and 2.6, which are known as *psychrometric charts.* These charts aid in our understanding of the composition of the air used in drying as it affects the rate of drying and as it affects processes within the ceramic during drying.

The starting point for understanding air–water vapor mixtures is to characterize them with respect to composition, that is, how much water vapor they contain. It is important to review Dalton's law from chemistry, which states that the total pressure of a mixture of gases is the sum of the partial pressures exerted by each gas in the mixture, each gas in the mixture exerts its own characteristic pressure independently of the pressures of other gases present, and the partial pressure of a gas in a mixture of gases is related to the total pressure by the mole fraction of the individual gases making up the mixture.

In an air–water vapor mixture, if p_T is the total pressure, p_W is the partial pressure of water vapor, and p_A is the partial pressure of air, then:

$$p_T = p_W + p_A \tag{2.8}$$

Applying the ideal gas law:

$$p_T = \frac{n_W RT}{V} + \frac{n_A RT}{V} \quad \text{(2.9)}$$

where: n_W = the number of moles of water vapor in the mixture.
n_A = the number of moles of air in the mixture.
R = the universal gas constant (0.08206 L-atm/mol-K or 10.73 ft^3-lb/in.^{2}/lb-mol-°R) (R = Rankin scale).
T = absolute temperature in K or °R.
V = volume of the confined atmosphere.

Using the data in Table 2.7 for the composition of dry air, the average molecular weight of air is calculated as 28.9645 g/mol. The molecular weight of water is well known to be 18.02 g/mol. Therefore, the molecular weight of an air–water vapor mixture (hereafter called a humid air mixture) is one way to scientifically characterize the composition of an air segment. The following definitions are useful in characterizing humid air segments from the perspective of mole fraction:

- *Vapor mole fraction:* the mole fraction of water vapor in an air segment.
- *Dry air mole fraction:* the mole fraction of constituents other than water vapor in an air segment. Since measurements of molecular weight require experimentation, other methods to characterize the composition of air segments are usually more convenient. The most convenient means is simply to express the composition as percentage by weight in a unit volume:

$$\%\ \text{water vapor} = \frac{\text{Weight of water}}{\text{Weight of water vapor} + \text{Weight of dry air}} \times 100 \quad \text{(2.10)}$$

- Parts per million by weight (ppm_w): Percentage water vapor by weight is an inconvenient quantity to use in calculations since the percentages are very small numbers. For example, air saturated with water at 20°C (70°F) contains only about 1.5% water vapor. A more convenient number is parts per million by weight, expressed as ppm_w, which is defined as:

$$ppm_w = \frac{\text{Weight of water vapor in an air mixture}}{\text{Weight of other constituents in the air (excluding water vapor)}} \times 10^6 \quad \text{(2.11)}$$

If the saturated air segment at 20°C is found to contain 1.5% water vapor by weight, then it contains 15,000 ppm of water vapor. An added advantage of expressing the humid air composition in this way is that ppm_w is a pressure- and temperature-insensitive quantity.

Just as the composition of a humid air segment can be expressed as percentage or ppm by weight, the composition can be expressed as percentage or ppm by volume. This requires consideration that in a unit volume of air, the composition is directly related to the partial pressure of the constituents or the number of moles present (through use of the ideal gas law).

- Percentage water vapor by volume: the ratio of the partial pressure (or volume) of water vapor to the total pressure (or volume) of or within the air segment (expressed as a percentage).
- Parts per million by volume (ppm_v): the ratio of the volume of water molecules per million molecules of other constituents in the gas segment.
- Mixing ratio by weight: ratio of the weight of water vapor to the weight of other constituents in an air segment.
- Mixing ratio by volume: ratio of the partial pressure of water vapor to the sum of partial pressures of other constituents in the air segment.

EXAMPLE 2.3: An air segment at 20°C is found to contain 4.54 g (0.01 lb) of water vapor per 454 g of dry air (1.0 lb). Express the composition of this air segment using all terms for air composition (except those using partial pressures) in Section 2.11.

a) Mole fraction:

$$4.54 \text{ g water} / 18.02 \text{ g/mol} = 0.807 \text{ mol}$$

$$454 \text{ g dry air} / 28.9645 \text{ g/mol} = 15.674 \text{ mol}$$

$$\text{Vapor mole fraction} = 0.807 / (15.674 + 0.807) = 0.049$$

$$\text{Dry air mole fraction} = 15.674 / (15.674 + 0.807) = 0.951$$

b) Percentage and ppm:

$$\text{wt\% water vapor} = 4.54 \text{ g} / (4.54 \text{ g} + 454 \text{ g}) \times 100 = 0.99$$

$$ppm_w = (4.54 \text{ g} / 454 \text{ g}) \times 10^6 = 10{,}000$$

c) Mixing ratio

$$\text{Mixing ratio by weight} = 4.54 \text{ g} / 454 \text{ g} = 0.01$$

Using the ideal gas law, the ppm_v is found to be 16,073 and the mixing ratio by volume is found to be 0.01607.

2.12 Absolute and Relative Humidity

Now that the composition of the atmosphere has been defined, other characteristics of the atmosphere useful in analyzing drying can be given. These are the well-known relative humidity of the air and other humidity terms used in technical operations. Further, the moisture content of air segments will be defined in language other than composition by employing *dew point* and *frost point.* Finally, the energy content of humid air segments will be considered.

The simplest definition of humidity for the average person is the sensation brought about by the water vapor content of the atmosphere at a given temperature. If the humidity is too low during winter months in cold climates, a person's skin may feel dry and simply walking across a carpet can generate static electricity. The weather report always gives the relative humidity as an indication of the water vapor content of the atmosphere.

In engineering practice, the most useful term is the absolute humidity *(H),* which is the weight of water per unit weight of a dry air. It is possible to express this weight in metric dimensions in kilograms of water per kilogram of dry air and in English units as pounds of water per pound of dry air. In heating and air conditioning calculations, the weight of water is frequently expressed as grains of moisture per pound of dry air where there are 7000 grains (gr) per pound. Using grains as a unit of weight is convenient because of the fact that the concentration of water in air segments near room temperature in pounds or kilograms is a small number leading to possible calculation errors. Because the convention in engineering calculations in North America is to use English units, absolute humidities in the rest of this text are expressed as pounds of water per pound of dry air or more conveniently as "lb/lb."

The absolute humidity is the weight of water vapor per unit weight of dry air in an air segment.

Relative humidity is the partial pressure of water vapor in an air segment expressed as a percentage of the partial pressure of water vapor at saturation, that is, a percentage of the pressure of water vapor where the air cannot carry more water.

The absolute humidity at saturation (H_{sat}), is the maximum water weight an air segment will carry. It is now possible to define percentage of humidity (H_p) and relative humidity (H_r). These are two different terms given the following definitions:

- *Percentage of humidity:* the weight of water in an air segment compared to the weight that the segment would carry if saturated (expressed as a percentage).

- *Relative humidity:* the partial pressure of water vapor in an air segment compared to the partial pressure of water vapor in the air segment when the air segment is saturated with water vapor (expressed as a percentage).

2.13 Characterizing Atmospheres Using Wet and Dry Bulb Temperatures

The classic way to determine relative humidity is to measure *dry bulb* and *wet bulb* temperatures. The relative humidity will be shown to be related to the difference between wet bulb and dry bulb temperatures (called the *wet bulb depression*). Many students are familiar with wet bulb temperature measurements because they have used a sling psychrometer in high school physics classes.

Dry bulb and ***wet bulb*** temperatures are used to characterize humid air segments.

The dry bulb temperature is the temperature obtained by inserting a measuring device into an air segment. Since these measurements have been typically made using a bulb-type thermometer, it is called a dry bulb temperature. It is the same temperature that would be obtained using a thermocouple or other sensing device that is simply inserted into the atmosphere. For convention in this text, the dry bulb temperature will be denoted as T.

The wet bulb temperature (T_w) is the temperature obtained by inserting a measuring device into an air segment when the measuring device or bulb is in constant contact with liquid water as might be supplied by a wick connected to a water reservoir. This traditional measuring technique is shown in Fig. 2.7.

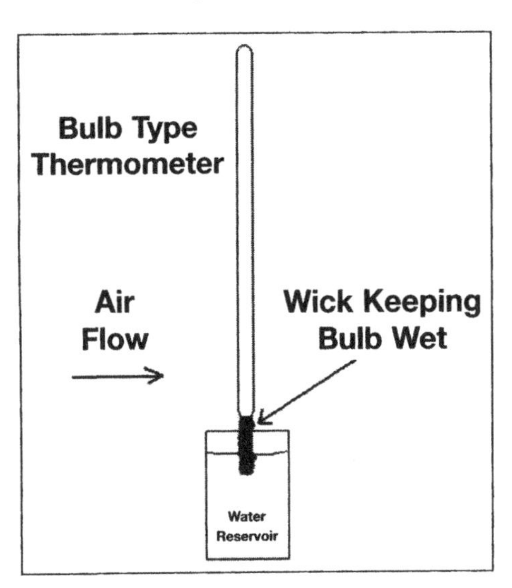

Figure 2.7: Wet bulb thermometer.

In considering the wet bulb thermometer, it is essential to realize that water freely evaporates from the wick around the bulb of the thermometer at a rate that depends on the relative humidity of the air flowing past the wick. The evaporating water absorbs heat from the wick and thermometer bulb, causing a depression or lowering of the sensed temperature. The amount of the wet bulb depression is then related to the relative humidity of the air segment.

To analyze the heat transfer

in the vicinity of the wet bulb, it is necessary to consider the heat transfer equation for convective heat flow in one dimension:

$$q = hA\Delta T \tag{2.12}$$

where: q = quantity of heat transferred per unit time.
h = film coefficient or proportionality constant for convective heat transfer.
A = the area exposed for evaporation.
ΔT = temperature difference between the wick and the air segment.

To apply Eq. 2.12 to the wick, it is simply necessary to show that heat flows away from the wick to the atmosphere as given by:

$$q = hA(T_A - T_w) \tag{2.13}$$

where T_A is the air dry bulb temperature and T_{wick} is the surface temperature of the wick covering the thermometer bulb. Of course, evaporation results in mass transfer (movement of water molecules) from the wick. If n_w is the number of moles of water vapor moving away from the wick, then n_w is found dependent on:

$$n_w = kA(H_{wick} - H_A) \tag{2.14}$$

where H_{wick} is the absolute humidity immediately at the wick surface and H_A is the absolute humidity of the air flowing past the wick. The term k is proportionality constant.

Finally, the heat *(q)* absorbed from the wick and bulb due to evaporation at the wick is:

$$q = n_w \Delta H_v \tag{2.15}$$

Equating Eqs. 2.13 and 2.15 while using Eq. 2.14, the result is:

$$T_A - T_w = \frac{k\Delta H_v}{h}(H_{wick} - H_A) \tag{2.16}$$

or more conveniently:

$$T - T_w = (\text{constant})(\text{saturation humidity} - H_A) \tag{2.17}$$

For any given dry bulb temperature, the saturation humidity is fixed, and the expression for the wet bulb depression becomes:

$$T - T_{\mathrm{w}} \propto H_{\mathrm{R}} \tag{2.18}$$

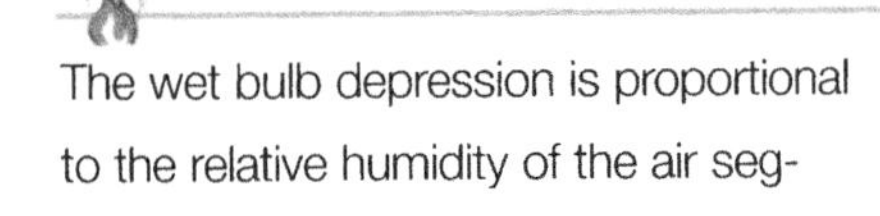

The wet bulb depression is proportional to the relative humidity of the air segment.

Relative humidity sensors were introduced into the ceramic industry in the 1990s. These sensors use humidity-dependent electrical properties of semiconducting compounds rather than wet bulb thermometers to measure relative humidity. These new electronic sensors have been shown to be sufficiently rugged to survive in ceramic dryers, making the days of wet bulb thermometers numbered. As will be shown in a subsequent chapter, when any two characteristics of an air segment are known, such as T and H_r, then all other properties of that air segment are known.

2.14 Enthalpy of Air–Water Mixtures

It was previously shown that water vapor significantly enhances the ability of air to carry and exchange energy. Since water vapor and air is a mixture, the enthalpy of that mixture, h, is:

$$h = (\text{Enthalpy of dry air}) + (\text{Enthalpy of water vapor}) + (\text{Enthalpy of mixing}) \tag{2.19}$$

or

$$h = h_{\text{dry air}} + (H_{\mathrm{r}})(h_{\text{sat}}) + h_{\text{mixing}} \tag{2.20}$$

For practical purposes, the enthalpy of mixing is negligible in magnitude compared to the other terms in Eq. 2.20. The enthalpy of air saturated with water in the low-temperature region is given in Table 2.12. Tabulated information on enthalpy of humid air mixtures is readily available and is given in a number of sources. These data have been used in many types of chemical processes and boiler applications where steam is used as a heat source.

In a like manner, computer software programs that are used with psychrometry furnish the enthalpy. For these reasons, and because enthalpy can be read directly from psychrometric charts (later chapters), it is not necessary to supply more data than are given in Table 2.12.

Table 2.12: Enthalpy of water-saturated air in the low-temperature region

Temperature (°C)	Enthalpy (cal/g)	Enthalpy (Btu/lb)
0	0 (reference point)	0
10	10.0402	18.074
20	20.0400	36.076
30	30.0244	54.049
40	40.0055	72.017
50	49.9896	89.990
60	59.9811	107.977
70	69.9839	125.984
80	80.0019	144.018
90	90.0395	162.088
100	100.1016	180.201

EXAMPLE 2.4: Compare the enthalpy that can be potentially recovered on cooling one pound of dry air from 80 to 20°C with that of one pound of saturated air cooled in the same interval.

From Table 2.11 (interpolating to obtain values):

Enthalpy of dry air at 80°C	152.0 Btu/lb
Enthalpy of dry air at 20°C	– 126.1 Btu/lb
Recoverable energy	25.9 Btu/lb

From Table 2.12:

Enthalpy of saturated air at 80°C	144.018 Btu/lb
Enthalpy of saturated air at 20°C	– 36.076 Btu/lb
Recoverable energy	107.9 Btu/lb

Condensation must occur on cooling the saturated air segment.

2.15 Dew Point Temperature and Frost Point Temperature

The dew point temperature, T_d, is the temperature at which an air segment exhibits condensation on cooling. Another way of stating this is that the dew point temperature is the temperature at which a given air segment is saturated and therefore cannot hold or carry more water.

With respect to drying of ceramic products, drying will not occur until the dew point temperature is exceeded. In the case of tunnel-type convection dryers, it is important to realize that condensation can occur in the entrance end of the dryer if the ware (or inspired air) reduces the air temperature below the dew point temperature.

Another way of considering the dew point temperature is to use a sports analogy. When playing golf in the early morning in humid climates, the grass will remain wet until the air temperature rises above the dew point temperature and drying begins. Less expert golfers need to play later in the day to take advantage of the extra roll possible when the grass is dry.

The ***dew point*** temperature is the temperature above which drying can take place within a given atmosphere. It is the temperature below which condensation can occur, possibly resulting in defective product.

The dew point temperature is one characteristic of a humid air mass that can be used to completely describe that air mass. However, since T and H_r (or T_w) are usually measured quantities during drying, dew point is usually calculated in order to avoid drying problems. It is also worthy of note that $T_d = T_w$ only when an air segment is saturated.

The frost point is the temperature to which a humid air segment must be cooled in order to begin condensing water vapor in the form of frost. The frost point exists only at temperatures below the freezing point of water.

2.16 Problems

1. While the water molecule is relatively stable, there have been instances in ceramic practice where breakdown of the molecule in the presence of carbon (as present in raw materials) has been cited as a reason for explosions in heated process equipment. Look up information on the steam–water gas reaction, and discuss conditions under which explosive atmospheres could be created in ceramic processing. Hint: Consider preheating in kilns.

2. A fired ceramic in a rectangular shape (9 × 4.5 × 3 in.) is placed on end in a shallow pan of water with the water level maintained at a constant depth and at room temperature (20°C or 70°F). The end of the shape is immersed only to a depth of about 0.5 in. After 24 h, the part is noticeably wet just above the water level with a uniform front of water visible 1 in. above the water level in the pan. If kerosene were used in this experiment rather than water for similar experimental conditions, what would be the approximate level of visible dampness above the level of the kerosene in the pan?

3. The experiment of Problem 2 with water is repeated at 40°C. What level does the

water reach in the ceramic part?

4. Use Eq. 2.7 to determine the saturation vapor pressure of water at 120°C.

5. Use the data in Table 2.5 and curve fitting techniques to find an equation for the saturation vapor pressure for water in the interval 0–100°C. Compare your result to Eq. 2.7, and explain any differences in form between your equation and Eq. 2.7.

6. Use equations in Section 2.12 and the data in Table 2.6 to calculate the relative humidity at saturation over a salt bath containing 3 mol of NaCl per liter of solution.

7. Why is powered aluminum metal pyrophoric in nature? (See Table 2.8).

8. An air segment at a dry bulb temperature of 90°F exhibits an absolute humidity of 0.0153 lb of moisture per pound of dry air. Determine the following characteristics of the humid air: mole fraction, percent water by weight, and ppm_w.

9. What is the relative humidity of the air segment in Problem 8?

10. (a) Why will calcium sulfate ($CaSO_4$) act as a desiccant, and what is its common or mineral name? (b) Discuss why calcium sulfate may fail to work as a desiccant after temperatures reach a certain level.

11. Using data in Table 2.12, consider cooling 10 g of saturated air from 80 to 50°C. How much water in grams condenses during the process based on the additional heat recovered over and above that simply due to the change in specific heat of the mass of air? Assume the latent heat of vaporization of water is constant between 50 and 100°C (Table 1.2).

CHAPTER 3

Drying Mechanisms in Particulate Systems

PART A: BASIC CONCEPTS

3.1 Introduction

The sequence of water addition to a completely dry ceramic was discussed in Chapter 1, Section 1.3, where water first covered the surface of the ceramic powder and subsequently filled fine capillaries and pores prior to causing a separation of particles. The sequence of water removal on drying was given as the reverse of the logical sequence of water addition; that is, the forming water is removed first (if present), followed in sequence by the pore water (if present), the capillary water, and the surface water.

In drying, moisture must move from the center of the formed piece to the surface where it can be carried away by the air mass (draft) in the dryer. Since transport is involved, it is of interest to discover the transport mechanisms so that the rate of transport (i.e., the rate of drying) can be controlled. It is also important to examine those factors that inhibit transport, such as attractive forces between the ceramic and the water.

Finally, since shrinkage is a usual consequence of water removal, it is of interest to look at the stress involved in drying. The goal of the discussions is to present strategies in drying from the perspective of the properties of the material.

Drying shrinkage is a consequence of water removal during drying.

3.2 Bound Water in Particulate Systems: Hygroscopic Water

Water molecules are attracted to ceramic surfaces because of the polar nature of the water molecule and its attraction for the surface of the ceramic, the latter of which bears an *apparent charge.* Most metal oxides bear an apparent negative charge at the surface because cations are typically surrounded by oxygen anions. Viewed from the outside, these particles appear as a negatively charged cloud.

Some ceramic particles exist as minute crystals bearing symmetry other than spherical or near spherical. Kaolinite crystals occur as hexagonal platelets whose flat surface (known as basal planes) are covered with oxygen anions giving them a negative apparent charge. By contrast, the edges of the platelets have an apparent positive charge due to the exposure of metal cations at the surface. Since water is polar, it has both positive and negative sides that can be attracted to the basal planes and to the edges of the crystals, respectively. Water is adsorbed on crystal edges as *broken bond water,* which is linked to unsatisfied charges on the surface. These adsorptions are illustrated in Fig. 3.1 where weak *Van der Waals bonds* are formed between the water molecules and kaolinite crystals.

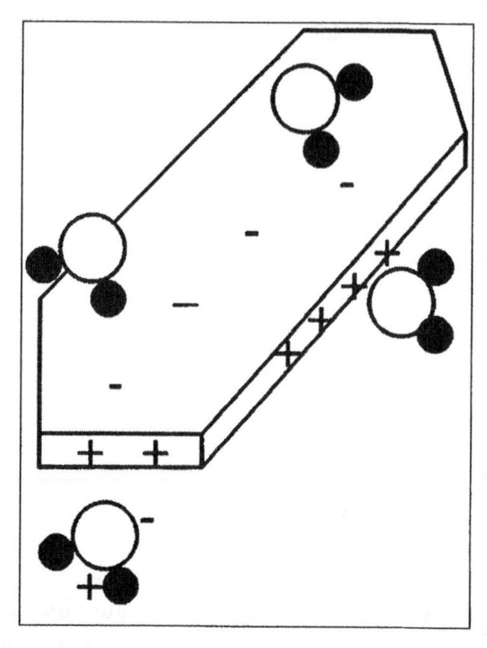

Figure 3.1: Attraction of water molecules to kaolinite surfaces.

Water that is weakly held to a particle surface is called *hygroscopic water.* This terminology is used since particle surfaces will attract water molecules, and water eventually covers the surface of exposed particles with a near monolayer. In fact, an equilibrium exists between an oxide powder and water vapor, with absorption increasing as relative humidity increases up to a maximum value — typically above about 40% relative humidity. Many polymer or cellular materials, such as wood and some organic species, continue absorbing water as humidity increases beyond about 40% relative humidity. Such behavior is illustrated in Fig. 3.2.

It is clear that the exposed surface area within the particulate mass affects moisture adsorption. In practical terms, it has been known for many years that different types of clays exhibit different equilibrium moisture adsorptions. In fact, moisture adsorption is used as an indirect indication of the mineralogy of clays. Some typical equilibrium moisture adsorptions are given in Table 3.1. The moisture adsorption is linked to the content of colloidal particles (those <~2 μm) in the various minerals.

> Moisture adsorption is a property of ceramic powders that is influenced by surface area and the chemical nature of the surfaces. Surface forces must be overcome to remove hygroscopic water in drying.

A common mistake in drying of ceramics is to assume that drying is com-

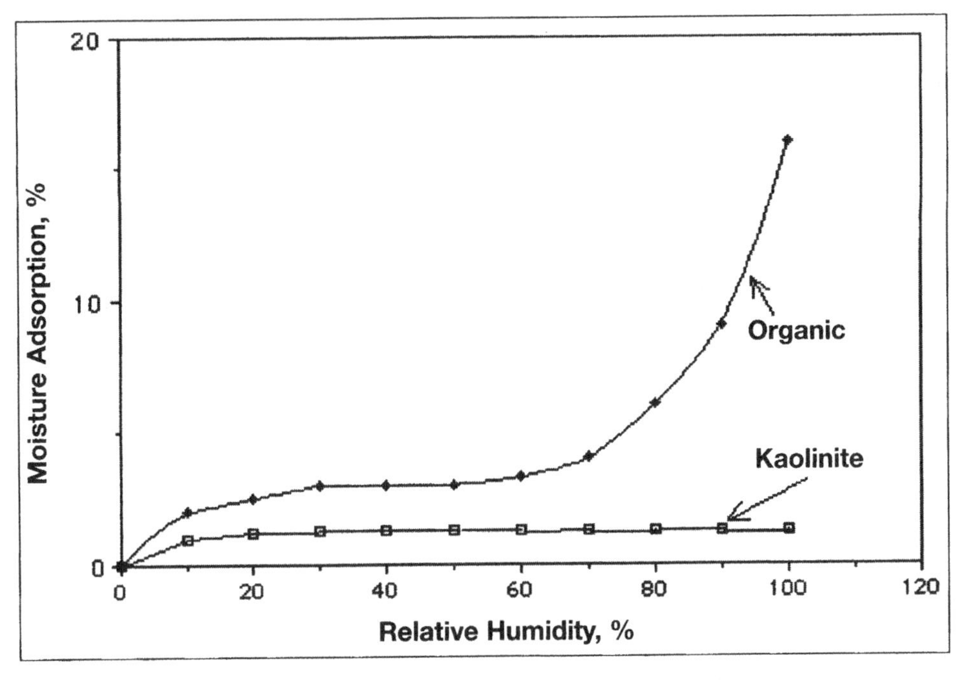

Figure 3.2: Moisture adsorption of ceramic and organic substances.

Table 3.1: Moisture adsorption of clay minerals

Mineral	Colloid content	Moisture adsorption (%)
Primary kaolinite	Low	~1.0–1.2%
Brick clay and fireclay	Low	~0.5–4%
Ball clay	Moderate	~5.5-7.5%
Bentonite	Very high	>7.5– ~14.5%+

pleted for all powders when the boiling point of water is exceeded. But, in fact, bentonite (Table 3.1) may require as much as 300°F (150°C) to be completely dried. This is because of the strong attractive forces between the bentonite particles and water molecules.

Another common mistake in drying of ceramics is the assumption that once the ceramic is dry, it will not reabsorb moisture from the atmosphere. However, it is well known that some dried brick products based on montmorillonite clays can reabsorb moisture from the atmosphere if they are allowed to dwell on a kiln car track prior to entering the kiln. Since montmorillonite can absorb large quantities of water (like the bentonite in Table 3.1), defective products can result from moisture readsorption due to expansion of the formed part during readsorption. The shrinkage during drying followed by slight expansions on readsorption can lead to formation of microcracks on the surface of the formed piece.

A simple laboratory test can be used to determine if a ceramic exhibits moisture readsorption (see Chapter 9, Section 9.17). The formed piece is dried to constant weight at a temperature of at least 150°C. The piece is removed from the dryer and placed in a room-temperature enclosure containing a salt solution that maintains a constant relative humidity (for example, a 6.0 mol NaCl aqueous solution, approximately 58 g of NaCl/l, will exhibit a relative humidity of ~77% at 20°C). The sample is exposed to the salt solution until it maintains a constant weight for at least 24 h. The moisture adsorption *(MA)* property is calculated as follows:

$$MA = \frac{(\text{Weight after exposure to constant humidity}) - (\text{Dry weight})}{\text{Dry weight}} \times 100 \quad \textbf{(3.1)}$$

3.3 Bound Water in Particulate Systems: Capillary Water

Water volume segments are drawn to and held in fine pores known as capillaries within packed particulate systems. Capillaries can be generally described as pores that have a diameter less than about 1–2 μm, about the same size as colloidal particles in the ceramic material.

Capillaries are pores formed as particles are compressed into close proximity in ceramic forming operations. It can be shown that the diameter of a pore between three spherical particles in contact is 0.154 times the radius of the spheres. In other words, if the average particle size of spherical particles is 1 μm, the average pore size is about 0.2 μm given perfect packing of the spheres. Since ceramic particles are almost never perfectly spherical, the actual pore size is some fraction of the particle size assuming the particles are of the same general size. Therefore, the capillaries are generally almost as large as the fine particles in the mix or raw materials used in manufacturing the product.

In reality a range of particle sizes is used in making ceramic products. Near perfect packing has been shown to result from a distribution of particle sizes when the distribution obeys a relationship similar to:

$$\frac{CPFT}{100} = \frac{d^n - d_s^n}{d_L^n - d_s^n} \quad \textbf{(3.2)}$$

where: $CPFT$ = cumulative percent finer than in the particle size distribution.
d = particle diameter at a particular $CPFT$.
d_L = the largest diameter particle in the distribution.
d_s = the smallest diameter particle in the distribution.
n = distribution exponent.

The relationship in Eq. 3.2 was derived by Dinger and Funk, who also defined a calculated quantity called *interparticulate spacing* (IPS). This latter quantity can help us visualize the size of pores in as-formed ceramics. For example, an extrusion of A-14 alumina (Alcoa) at 20.2% moisture content (DB) was found to exhibit an IPS of 0.435 μm. Since the median particle size of the A-14 was on the order of 2 μm, the fluid layer between particles in the undried (green) ceramic is of a size considered to be a capillary.

Usually larger voids are found in particle size distributions for either of two reasons: imperfect packing with voids extending over a range of particle diameters, or intentional or unintentional gaps in the particle size distribution.

In ceramic products produced by dry pressing, *gap graded* sizing can be employed so that there is a deficiency of medium-sized particles (in the general range 0.15–1.5 mm) and an excess of coarse particles (>1.5 mm) and fine particles (<0.15 mm). In such cases, the pore size distribution is found to be *bimodal* with pores generally appearing in size range of the smallest particle size and in the area of 70–120 μm. The smaller pores in the distribution are generally smaller than about 10 μm, and many of these are in the capillary size range. Therefore, most green (unfired) and many fired ceramic products contain capillary-sized pores.

A consequence of the presence of capillaries in green (or dried) and fired (porous) products is that the products will attract moisture (over and above hygroscopic moisture). One experiment to demonstrate this property is to take a fired ceramic product and place it in a pan containing water so that the part (formed ceramic) is only partly immersed. The water will be drawn into the ceramic and a liquid front of water will be seen slowly rising along the side of the ceramic. The property of the ceramic causing the absorption and movement of water is called *capillarity.*

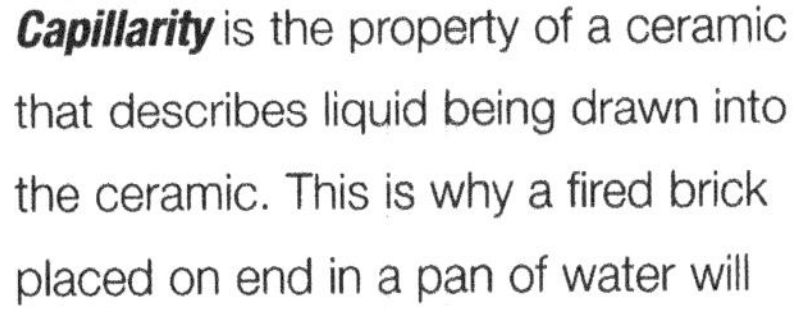

Capillarity is the property of a ceramic that describes liquid being drawn into the ceramic. This is why a fired brick placed on end in a pan of water will draw the water up into the brick in apparent defiance of gravity.

Capillarity can be conveniently measured in such a simple experiment as described above. It is usually found that the rise of water will be several centimeters over a period such as 24 h. In the case of dried products, capillarity can be measured using kerosene as the liquid to be drawn into the ceramic. Using water results in slaking and collapse of most dried ceramic parts. Capillarity is usually quantified as the weight gain due to capillary suction at a specified immersion time. See Chapter 2, Section 2.4, for a treatment of capillary suction.

From the discussion above, capillaries are shown to attract and hold water. Therefore, during movement of water in drying capillaries will tend to resist being drained or depleted. However, a competing effect to increase removal of water in capillaries is found with respect to vapor pressure over a capillary, p_{wc}, where it can be shown:

$$\ln\left(\frac{p_{wc}}{p_{wf}}\right) = \frac{2\gamma m}{RT\rho}\left(\frac{1}{r}\right) \tag{3.3}$$

where: p_{wf} is the vapor pressure of water over a flat surface.
m is the molecular weight of the fluid in the capillary.
R is the gas constant.
T is the absolute temperature.
r is the capillary radius.

This means the vapor pressure of water increases dramatically as the radius of the capillary decreases at constant temperature, that is:

$$p_{wc} \propto \exp\left(\frac{1}{r}\right) \tag{3.4}$$

Equation 3.5 below and inference from Eq. 3.4 imply that evaporation from the surface of the capillary is the most rapid means of removal of water since the capillary resists drainage by another means.

3.4 Nonbound Water in Particulate Systems: Pore Water and Forming Water

Water in larger pores (diameter >10 μm) can move with reasonable freedom through the ceramic during drying in a manner similar to how water moves through a filter — particles in the ceramic impede the flow but do not attract the water molecules since the surface requirement has already been met. Under these conditions, the flow from a region of high pressure, p_2, to a region of lower pressure, p_1, is described by Darcy's equation for flow in porous media:

$$\frac{p_2 - p_1}{L} = \frac{\alpha\mu V}{g_c} \tag{3.5}$$

where: L = path length though the porous medium.
α = viscous resistance coefficient in ft^{-2}.
μ = fluid viscosity.
V = velocity of flow of the fluid.
g_c = dimensional constant (32.17 lb-ft/lb-force/s^2).

While Darcy's equation adequately describes movement of water within the actual core of the ceramic during drying, it is found that the movement of water vapor to the surface of the part is the rate-limiting or rate-controlling step. The movement of water vapor will be treated as a diffusion process in a later section.

Darcy's equation describes flow of water in larger pores during drying. The rate of movement of water is controlled by the resistance to flow by factors such as the amount of porosity and how the shape of pores impedes flow; i.e., the permeability of the green body.

After larger pores have been filled with water during mixing, additional water results in a separation of particles. In most cases involving plastic forming of ceramics (e.g., extrusion or plastic pressing), the apparent viscosity of the wet ceramic (sometimes called a *paste*) decreases very rapidly as the percentage of water increases beyond that required to fill the larger pores. It is beyond the scope of this text to treat the subject of rheology; however, the point is made that if additional water results in separation of particles, then removal of that water brings the particles closer together. *It is the forming water that results in shrinkage.*

PART B: STAGES OF DRYING

3.5 The Rate and Stages of Drying

The *generalized rate of drying* of a ceramic product is shown in Fig. 3.3. The rate of drying is conveniently defined as the change in the weight of moisture in the ceramic per unit time. In a plot of the moisture weight in the ceramic versus time, the rate of drying is then the slope of the line defined by the drying process. It is clear that the slope of the line is uniform or constant during the initial drying of the ceramic, for example, on progression from Point A to Point B in the drying process of Fig. 3.3. For this reason, the initial period is called the *constant rate period* of drying.

THE PROGRESSION OF DRYING IS:

- The constant rate period.
- The first declining rate period.
- The second declining rate period.

As drying continues beyond Point B, the drying rate declines with the rate of decline constant. In other words, the slope of the line in Fig. 3.3 as drying progresses from Point B to Point C is declining at a consistent rate. The period of drying in progressing from Point B to Point C is called the *first declining rate period.* In the next period of drying (progression from Point C to Point D) the rate of drying declines at an increasing rate; this period is called the *second declining rate period.*

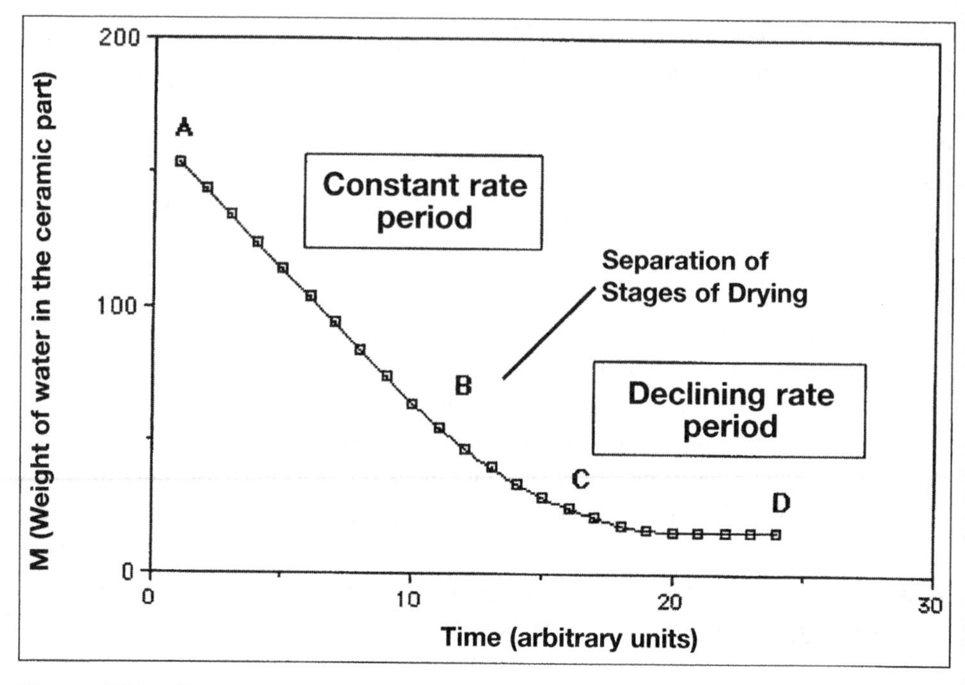

Figure 3.3: Change in moisture content in ceramics during drying.

The information presented in Fig. 3.3 can be more conveniently presented in a plot of drying rate versus time. As shown in Fig. 3.4, at the start of drying the rate of drying is constant on progression from Point A to Point B. Beyond Point B, the rate of drying declines constantly until Point C is reached. Then, the drying rate progressively slows until drying is completed. Figures 3.3 and 3.4 present the same information, but Fig. 3.4 more dramatically illustrates the distinct periods in drying ceramic products.

It is important to divide the drying process into two stages. Stage I is the constant rate period, and Stage II contains both the first and second declining rate periods. As will be shown later, the moisture content at Point B in Figs. 3.3 and 3.4 is called the *critical moisture content* (M_c) because shrinkage takes place as drying continues down to M_c, but shrinkage is negligible below M_c. When clay ceramics reach M_c, they are called *leather hard.* In traditional ceramics, this is the point at which the product can be moved without fear of warpage.

DRYING IS DIVIDED INTO TWO STAGES:

- Stage I: The constant rate period during which shrinkage is occurring. When M_c is reached, shrinkage stops.
- Stage II: The first and second declining rate periods.

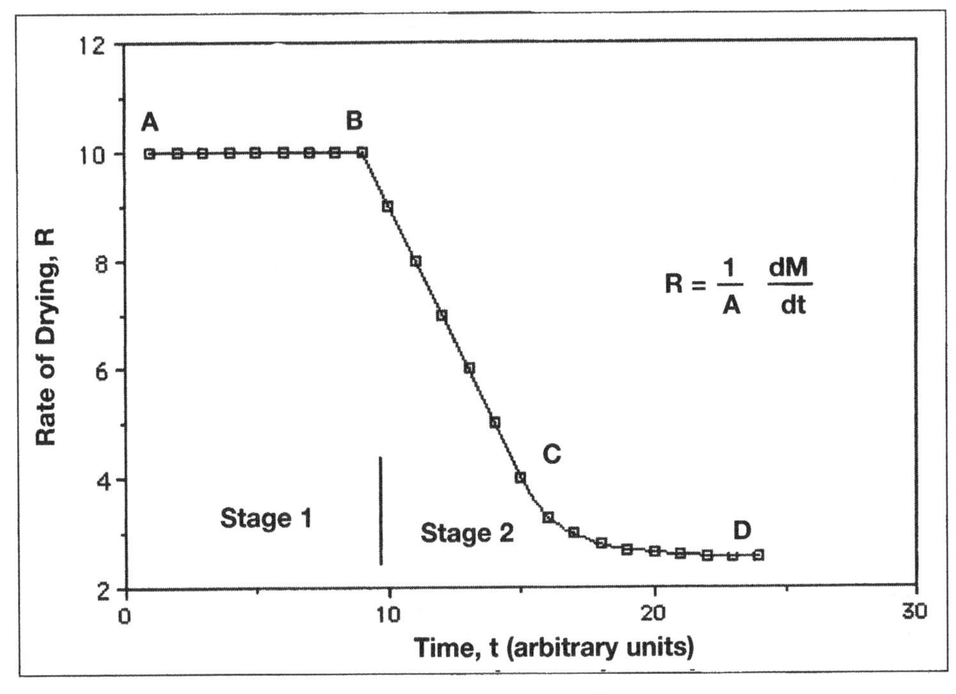

Figure 3.4: Drying rate plot. M = weight of water in the ceramic; A = surface area of the ceramic exposed to the drying environment.

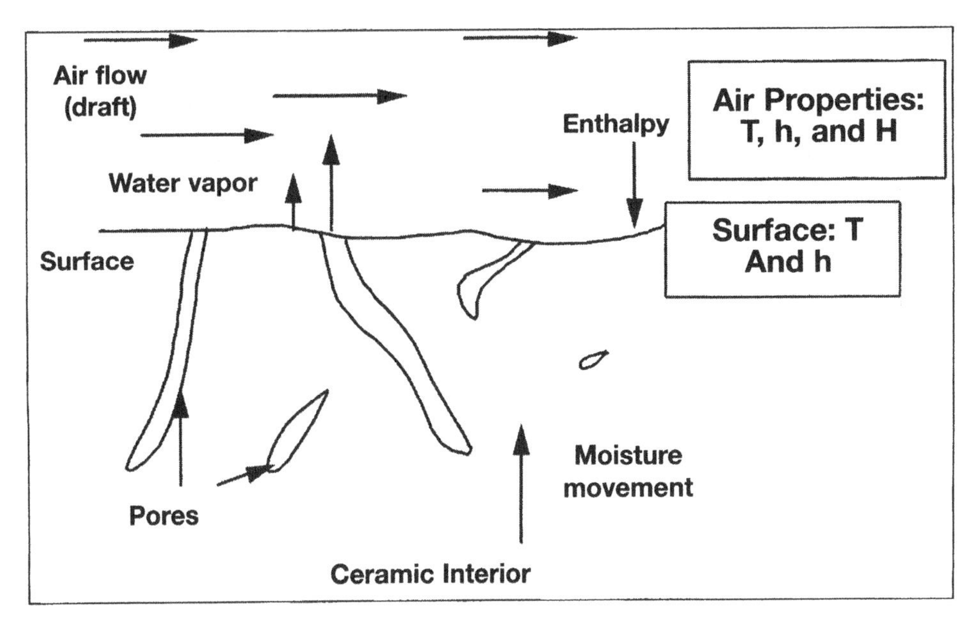

Figure 3.5: Model of cross section of a drying ceramic.

3.6 Stage I: The Constant Rate Period

The fundamental assumption in the constant rate period is that the surface of the green ceramic part is constantly covered with a continuous film of water. As evaporation takes place, the water on the surface is continuously replenished by capillary suction acting on the interior of the ceramic. This mechanism of replenishment of the water on the surface is generally called *wicking*. The rate of evaporation is the same as from a free or exposed surface of water that would be held in a container and exposed to the same conditions of temperature, ambient humidity, and ambient air velocity.

The mechanism of drying in Stage I is surface evaporation and moisture movement by capillary suction called ***wicking.***

The model for this process is shown in Fig. 3.5, where the cross section of a segment of the ceramic is shown with the airflow in the dryer interacting with the surface. The properties of the air segments in the bulk of the draft (*T*, *h*, and *H*) are shown along with the properties of the air segment directly in contact with the surface.

If all heat transfer at the surface of the ceramic is by convection, then:

- The surface temperature is the wet bulb temperature of the air segment flowing over the ceramic.
- The enthalpy of the surface is the enthalpy established by the wet bulb temperature.

The rate of evaporation in drying, *R*, is the rate of water removed in weight of water per unit area of exposed surface in the water (lb/ft^2 or kg/m^2). The rate is proportional to the enthalpy difference between the surface (h_{surf}) and the bulk of the air segment h_{amb}:

$$R = \zeta\left(h_{\text{surf}} - h_{\text{amb}}\right) \tag{3.6}$$

where ζ is a proportionality constant called a *mass transfer coefficient.*

The rate of evaporation is also governed by convective heat transfer, as it was in Eq. 2.26. So:

$$R = \frac{h\left(T - T_{\text{w}}\right)}{\Delta h_{\text{vap}}} \tag{3.7}$$

where h is the film coefficient for convective heat transfer and Δh_{vap} is the heat of vaporization of water. The terms T and T_w denote dry bulb and wet bulb temperatures, respectively.

The film coefficient h is found to vary with air velocity. Various equations are employed to describe the variation in h, and further information is available in references. In the simplest case, h is described as follows:

- For air flow parallel to a surface:

$$h = 0.0128G^{0.8} \tag{3.8}$$

  where G is the mass velocity of the air in lb/ft^2-h. The dimensions of h are Btu/h-ft^2-°F.
- For air flow perpendicular to a surface:

$$h = 0.037G^{0.37} \tag{3.9}$$

The information in Eq. 3.7 indicates the rate determinants for drying in Stage I. These are the difference between the dry and wet bulb temperatures (i.e., the wet bulb depression) and the air flow conditions allowing for heat transfer and transport to water vapor away from the surface where evaporation takes place.

It is important to note that temperature of drying air alone is not the only driving force for drying. It is obvious that there must be sufficient sensible heat to meet the requirements for evaporation, but the driving force for evaporation is also linked to the ability of the air to carry water vapor away from the surface. Further ways of illustrating this point are:

The drying in Stage I is controlled by the wet bulb depression (i.e., the ability of the air to carry water) and the heat transfer conditions that supply enthalpy for evaporation. Excessive sensible enthalpy in drying air can cause non-steady-state conditions in drying, resulting in steam explosions within the product.

- Drying will not occur if the air surrounding the ceramic part is saturated with water vapor regardless of the dry bulb temperature of the air.
- In drying in Stage I, the surface temperature cannot exceed the boiling point of water in most situations. This is analogous to the water in a kettle on a stove. The water in the kettle will remain at 100°C while it is boiling (for 1 atmosphere total pressure). It is only when all of the water has boiled away that the teapot can exceed 100°C.

- If heat transfer exceeds the conditions for normal drying in Stage I, excessive steam will be generated within the body, causing a steam explosion. For this reason, the air used in drying must not contain too much sensible heat; that is, the dry bulb temperature must not be too high.

Since shrinkage is occurring in Stage I, the rate of drying in Stage I must be limited to prevent cracking. In a later section, a *maximum safe drying rate* will be developed. It is worthy of note that once Stage I ends (i.e., M_c is achieved), the drying strategy changes since shrinkage is not longer a critical consideration.

It is important to consider that evaporation is occurring only at the surface with water transported or wicked to the surface of the part. If salts within the ceramic composition have dissolved, these will be transported to the surface along with the water, and the salts will be deposited at the surface as a consequence of evaporation.

Scumming is a phenomenon of discoloration or fluxing on a fired surface resulting from deposition of salts on the surface during drying.

3.7 Stage II: The First Declining Rate Period

After M_c is reached in drying, the first declining or falling rate period begins. The surface is no longer covered with water, and evaporation begins below the surface of the ceramic part. The dominant transport mechanism is capillary suction or wicking.

The suction pressure is created by evaporation of water from the capillary surfaces. As water evaporates from the surface, the capillary attempts to drain the interior of the green part to maintain water in the capillary, creating suction.

There are two controlling variables in this period:

- Dry bulb temperature of the drying atmosphere as well as heat transfer rates.
- Ability of the ceramic to expel water vapor (rate of permeation of water vapor).

The primary role of temperature in the first declining rate period is to increase the permeation and diffusion rate of water vapor through the pore structure of the dried periphery of the product. It is true that other effects are operative. For example, the viscosity of water in the capillaries continues to decrease, affecting the transport rate of water in the capillaries. However, it is the diffusion rate within the dried segment of the part that is controlling the rate. For this reason, it is commonplace to supply additional temperature to the product in Stage II to accelerate drying.

3.8 Stage II: The Second Declining Rate Period

In the second declining rate period, the residual water in capillaries and the surface water is removed. The primary mechanism of drying is diffusion of water vapor out of the product. Since diffusion controls the rate, temperature in the dryer is the main control variable, and temperatures in this stage are the highest during the drying process.

Temperature is increased during Stage II to increase the drying rate, as there is less probability of damaging the product as compared to Stage I.

There is no exact line of demarcation between the first and second declining rate periods except that the slope of the drying rate curve begins to decline in a manner suggesting a *parabolic relationship*. In fact, many transport-controlled processes observed in ceramic materials obey a *parabolic rate law* of the form:

$$y^2 = kDt \tag{3.10}$$

where: y = dependent variable (such as thickness of a drying product's dried outer layer).
k = proportionality constant.
D = diffusion coefficient.
t = time.

If the rate of drying, R, can be expressed as the differential relationship dy/dt, then:

$$R = \frac{dy}{dt} = \frac{\left(\frac{k}{2}\right)D}{y} \tag{3.11}$$

Equation 3.11 predicts that the rate of drying decreases continuously as the process of drying progresses during the second declining rate period (as y increases). This is, in fact, the shape of the drying rate curve of Fig. 3.4 (the progress from Point C to Point D).

Since diffusion is a thermally activated process, the diffusion coefficient follows an Arrhenius-type relationship, such as:

$$D = D_0 \exp\left(\frac{-Q}{RT}\right) \tag{3.12}$$

where: D_0 = proportionality constant.
Q = activation energy for diffusion.
R = gas constant.
T = absolute temperature.

What this simply means is that as temperature is increased, the rate of diffusion of water molecules in drying increases exponentially. In the second declining rate period, high temperatures are required to complete the drying process in a reasonable time frame.

Temperature and heat transfer are the primary control variables in Stage II of drying because they affect the rate of diffusion of water molecules that accomplish drying.

3.9 Introduction to Drying Shrinkage

Drying shrinkage results as water is removed in Stage I of drying because removal of water allows particles to collapse toward each other acting in response to capillary forces. These forces are a result of the overlap of water films. Drying processes affect the ultimate properties of the ceramic when defects are created, or perhaps revealed, during drying. The primary defect is crack formation (discussed in Chapter 7).

Drying must take place due to evaporation from the surface of the formed part. This means the surface shrinks before the interior of the part, creating *differential shrinkage* between the exterior and the interior. Because the outside of the part is shrinking faster than the interior, the surface of the part is placed in tension. Due to the low strength of the parts prior to firing, cracks can be initiated.

Differential shrinkage can also take place as a consequence of operational factors such as nonuniformity created in forming and uneven airflow during drying. The simplest example of the latter is a part of significant dimension on the supporting surface placed on an impermeable support (such as a steel tray) during drying. In this case, the bottom of the ceramic dries more slowly than the top, leading to cracks due to lack of access of drying air to all sides of the part. The simple remedy is to place parts on perforated trays.

Throughout the history of ceramic processing, it has been known that there are three primary factors in avoiding defects in drying:

- The rate of drying during Stage I must be sufficiently slow to avoid stress development, which initiates cracks.
- The moisture content of the ceramic mix affects drying shrinkage.
- The particle size distribution of the ceramic material has a significant effect on the drying shrinkage that is observed.

The rate of drying during Stage I is of paramount importance in producing crack-free parts. Because of this, the Stage I drying cycle typically employs the lowest drying temperatures and most humid atmospheres to control the drying rate. With the current status of ceramic technology, it is not possible to precisely predict maximum safe drying rates (also called *critical drying rates*). Practical experiments in humidity-controlled laboratory dryers are conducted to determine the critical drying rate for a particular product.

The critical drying rate is the maximum drying rate for a material without cracking due to shrinkage in Stage I.

The importance of the initial drying period can be illustrated by considering techniques used by potters in producing ceramic articles. Immediately after the piece is completed, it may be taken to a draft-free area and covered by wet cloth to maintain high humidity over the piece. As the cloth gradually dries, the piece also dries, but at a slow and uniform rate. After a period of a few days, the piece may be uncovered so that drying can continue. Whether potters realize it or not, the piece probably reaches critical moisture content while covered, so the drying rate is much less important in the ensuing period. Finally, the piece may be placed in an oven to accelerate drying.

In industrial processes, the initial drying period may take place in a dryer or in a separate process known as *predrying*. Predryers typically employ air at less than 100°F (<40°C) and allow for high relative humidity (H_r > 80%). The amount of moisture removed may depend on the need for subsequent forming operations. In turned ceramic products (those formed by jiggering, a plastic forming operation), predrying may involve an initial moisture loss of only about 2–3% so that the product can be handled without deformation. Where subsequent forming is unnecessary, as in production of bricks, predrying can involve removal of all forming water so that M_c is reached. In this case, all of the shrinkage took place in predrying.

Viewed on a particulate scale, drying in Stage I involves the inward collapse of particles. In relatively slow drying processes, particles are free to move on a limited scale and exhibit limited translation and rotation. If drying is slow, particles can move in a manner that reduces stresses created by shrinkage. In the event of fast drying, particles come into contact for less time and are not free to move. In such instances, cracks are more likely to develop.

Slow drying rates in Stage I allow particle movements that reduce stresses caused by shrinkage.

The moisture content of the ceramic mix influences drying shrinkage by causing further average separation between particles as water content increases. In ceramics produced by plastic forming, slight increases in water content cause significant increases in plasticity, producing softer and more workable materials with higher drying shrinkage. Data for a brick raw material are given in Table 3.2.

The particle size distribution has a primary influence on drying shrinkage. It

Table 3.2: Moisture content related to forming method and drying shrinkage with a brick raw material

Forming method	Moisture content (%WB)	Penetrometer reading*	Linear drying shrinkage (%)
Stiff extrusion	16.0	4–5	2.5–3.5
Soft extrusion	19.0	1–1.5	3–4
Molding	24.0	NA	4+

**kg force to create a ~4 mm diameter impression to a depth of ~4 mm at the surface*

has long been realized that nonplastic particulate additions, commonly called *grog*, reduce drying shrinkage. The addition of coarse particles (>100 μm diameter) is not possible in many ceramic products because of surface finishing requirements, but in products where it can be used, it remains a solution to excessive drying shrinkage given required production rates.

An age-old practice in ceramics formed by plastic processing is to add grog to reduce drying shrinkage.

Grog particles may range in size up to about 3 mm diameter, but surface finish and cutting processes impose limits on grog size. Grog materials include sand, natural stones, ash, calcined aggregates, and sawdust. Grog is usually used when a predominance of particles in the raw material are in the clay size range — that is, <2 μm.

In refractories and technical ceramics, size-graded raw materials are used to provide for closely controlled density and physical properties in the fired product. In these instances, unexpected changes in particle size distributions can lead to drying cracks. It has been said, "An engineer who controls his/her particles also controls his/her business."

Some ceramic materials exhibit *dewatering* during forming under pressure. Dewatering is seen when parts exhibit variable moisture content despite practically uniform mixing. An example is a ceramic produced in a piston extruder where the first product out of the extruder exhibits a moisture content lower then the product made near the end of the stroke of the extruder. What is happening is that during compaction in the extrusion die, water is forced under pressure in a direction opposite to the travel of the piston (direction of extrusion). This means the mix is unable to uniformly carry the water through the die, indicating a need for a different particle size distribution and binder/plasticizer addition. Those parts extruded last have a higher moisture content and an increased probability of cracking. The cause of the problem was dewatering in the extruder.

Grog is used in many ceramic mixes to reduce drying shrinkage and avoid drying cracks.

Many problems leading to crack formation on drying are attributed to forming processes. The cracks appear during drying, implying that there was a drying problem. More information on these situations is given in Chapter 7.

3.10 Problems

1. A clay specimen is held in an aluminum weighing cup and is dried to constant weight at 150°C. The specimen weight is 12.236 g and the cup weighs 6.255 g. The dried specimen is held over a closed salt bath at 20°C that contains 2.0 mol of NaCl per liter of water. After 24 h over the salt bath, the cup and specimen weigh 19.256 g.

   (a) What mineralogy is suggested for the clay material (by using Table 3.1)?

   (b) Using the composition of the salt bath and information in Fig. 3.2, would this clay specimen likely adsorb more water in a longer dwell over the salt bath?

   (c) Provide an exact method for determining clay mineralogy.

2. The extrusion pressure for a mixture of calcined alumina increased dramatically as the particle size distribution changed from an n value of 0.4 to an n value of 0.2 (Eq. 3.2). The extrudate or formed parts increased in density as the n value decreased.

   (a) How did the particle size distribution change as reflected by the change in n?

   (b) What is the implication for possible drying problems as reflected by the change in extrusion pressure?

3. Using Darcy's equation, calculate the relative velocity of water in a porous medium at 60°C compared to 20°C (using data in Table 2.4).

4. Data on moisture content of a ceramic part (% WB) versus time for a ceramic is given below:

Time (h)	Moisture content (%)
0	8.5
2.5	6.0
4.5	4.0
5.0	3.5
6.0	2.75
7.0	2.10
8.0	1.55
9.0	1.20
10.0	1.15

(a) Plot moisture content versus time for the data.

(b) Where is M_c for this material?

(c) What is the drying rate per unit area during the constant rate period where A is the surface area of the part (provide an equation)?

5. Contrast the rate of drying in Stage I for a ceramic when the dry bulb temperature is 212°F and the wet bulb temperature is 190°F with the mass velocity of the air flow as 2.0 lb/ft^2-h given:

(a) Air flow parallel to the surface of the drying ceramic part.

(b) Air flow perpendicular to the drying ceramic part.

6. The pore size distribution is measured on an extruded clay ceramic and on another extruded ceramic made using the same clay but with 5 wt% of a –8+20 mesh fired grog added. The pore size distributions of both dried products are measured using a mercury intrusion porosimeter with the following results:

Porosity data	100% clay	95% clay, 5% grog
Total porosity (%; from total Hg intrusion data)	16.0	16.5
% of pores <2 μm diameter	98.0	91.7
% of pores >20 μm diameter	1.0	5.5

(a) Draw a picture representing a cross section of the dry ceramic, showing how grog particles might be arranged to create larger pores.

(b) Why is it said that grog particles can create microcracks during drying that facilitate release of water vapor?

7. The particle size distribution of plastic (moldable) and castable (refractory concrete) refractory ceramics can follow a particle size distribution such as given below:

% retained on U.S. sieves

Sieve	Opening (mm)	Plastic (moldable) refractory	Castable refractory
4	4.76	2.4	5.5
8	2.38	6.8	17.6
12	1.68	8.6	12.0
20	0.84	6.6	8.8
30	0.59	5.5	7.9
50	0.297	7.0	8.2
100	0.105	18.9	9.5
200	0.074	22.4	12.8
Pan	0	21.8	17.7

The plastic refractory is extruded and shipped to the job site where it is placed by pneumatic ramming. It is subsequently dried before use at high temperatures. The castable is shipped to the field dry (as it contains hydraulically setting calcium aluminate cement) where it is mixed with water and it may be placed by casting behind forms. The castable must be cured and subsequently dried prior to exposure to high temperature.

(a) What can happen in field drying of the plastic if a batching error is made at the refractory plant where the total coarse material (fractions on 4, 8, and 12 mesh) is inadvertently reduced to 8.0% of the batch with the balance as –50 mesh material?

(b) Usually plastic refractory is dried within 3 days after placement in construction. In one case, a large furnace with a flat suspended roof of plastic (supported by anchors) was dried six months after placement. The plastic exhibited 1.5% linear shrinkage on drying under laboratory conditions. However, in the case of the large furnace with delayed drying, large cracks opened after the eventual dryout. Why did slow air drying allow cracks to open in an exaggerated fashion in contrast to normal practice, where cracking is minimal?

(c) In dryout of castable refractory, both placement moisture (allowing flow during casting) and moisture involved with cement hydrate phases must be released. What can happen if excessive preheat rates are used in dryout?

(d) In order to accelerate drying of some castable refractories, polypropylene fibers are added to act as conduits to facilitate steam release. Both solid and hollow fibers are in many cases effective. How can a solid fiber act to enhance steam release in drying?

8. In Stage II of drying, the rate of drying is controlled by the rate at which diffusion takes place. Name two factors related (a) to the dryer operation and (b) to the material design that can affect the diffusion coefficient and thereby affect the rate of drying.

9. Shelling is the loss of surface layers from a drying ceramic due to explosive spalling (steam spallation) during Stage II of drying. In drying brick, the defect may be a surface layer 0.25–0.5 in. thick that covers part or all of the exposed face of the brick during drying. If shelling is observed and is isolated to the bottom area of the setting of the product on kiln cars, what is the likely cause of the problem?

3.11 References

Dennis R. Dinger and James E. Funk, *Predictive Process Control of Crowded Particulate Suspensions Applied to Ceramic Manufacturing.* Kluwer Academic Publishers, Boston, 1994.

John H. Perry, C. H. Chilton, and S. D. Kirkpatrick, *Chemical Engineer's Handbook,* 4th ed. McGraw-Hill, New York, 1963.

CHAPTER 4

Psychrometry

PART A: READING AND INTERPRETING PSYCHROMETRIC CHARTS

4.1 Basic Concepts

Psychrometry is the science of air and vapor mixtures; specifically, it is the characterization of an air–vapor mixture in terms of basic properties of that mixture. The basic properties of interest with humid atmospheres include the dry bulb temperature, the wet bulb temperature, the relative humidity, and the enthalpy of the mixture. Because engineers are concerned primarily with evaporating water in drying of ceramics, psychrometry as applied to drying of ceramics is the science of air and water vapor mixtures utilized to understand and control dryers.

Psychrometry is a necessary element in interpreting processes within dryers; it is the viewpoint of drying from the outside of the ceramic part or component. The viewpoint of the ceramic from within the ceramic, involving moisture movement and stages of drying, was discussed in Chapter 3. Now, given the constraints of a particulate system, it is necessary to investigate drying from the standpoint of the dryers. In order to understand what is happening within the dryer and to control the process of drying, psychrometry is employed.

Psychrometry is the characterization of air–water vapor mixtures. The tools of psychrometry allow drying to be viewed in terms of the environment of the ceramic part during drying.

Psychrometric charts are used to analyze drying situations. In Chapter 2, the vapor pressure of water was related to temperature with a saturation boundary established as the maximum amount of water that air was capable of absorbing or carrying at a given temperature (Fig. 2.5). This saturation boundary is also a phase boundary, because crossing this boundary (i.e., taking an air segment across the saturation line) produces a supersaturated condition and condensation must occur (change

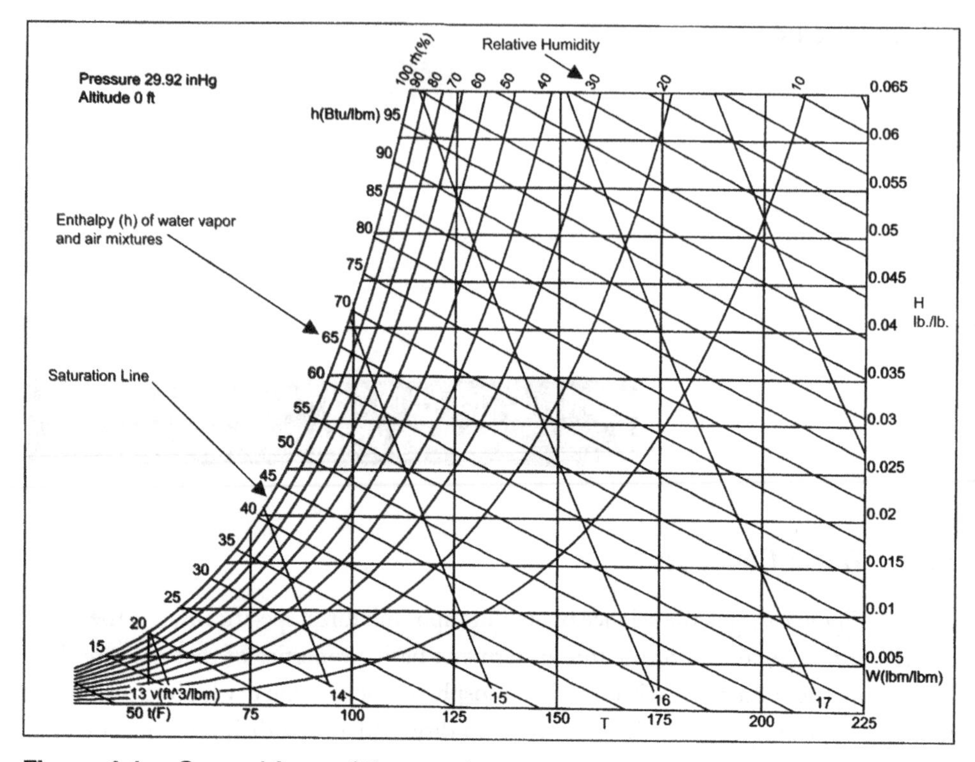

Figure 4.1: General form of the psychrometric chart.

from the vapor to the liquid state). Condensation continues in this case until the saturation vapor pressure is reachieved.

A simple psychrometric chart for an air–water vapor mixture at 1 atm total pressure is shown in Fig. 4.1. It is fundamentally a plot of absolute humidity versus temperature. On first viewing of this chart, most students believe that it is very complicated. However, comparison of Fig. 4.1 to Figs. 2.5 and 2.6 shows that the psychrometric chart is simply a more detailed version of the saturation relationship for water vapor or a segment of the phase diagram for water. The saturation line is plotted using the Clausius-Clapeyron equation (Chapter 1) to define the phase boundary. The equation can be modified to plot families of vapor pressures (or absolute humidities) at differing relative humidities, to allow calculation of enthalpies, and to determine other psychrometric properties of interest (this chapter, Part B). This allows the entire chart to be generated.

The area under the saturation line in Fig. 4.1 includes all humid air mixtures that are not saturated. Drying is possible using air of characteristics of all possible points under the saturation line since that air can receive or carry additional moisture.

Table 4.1: Summary of humid air properties

Symbol	Description	Dimension (English units)
H	Absolute humidity – water vapor content of a humid atmosphere on a dry air weight basis (also denoted as W on charts in Chapters 4 and 5)	lb/lb dry air or lb/lb
H_r	Relative humidity, a percentage of water vapor in air relative to the water vapor content at saturation	%
T	Dry bulb temperature	°F
T_w	Wet bulb temperature	°F
T_d	Dew point temperature	°F
h	Enthalpy of a humid air mixture on a dry air weight basis	Btu/lb of dry air or Btu/lb
V	Volume of an air segment per unit weight of air	ft^3/lb
V_d	Volume of an air segment on a dry air weight basis	ft^3/lb of dry air or ft^3/lb
V_h	Volume of an air segment on a wet air weight basis	ft^3/lb of moist air or ft^3/lb

Any air segment on the chart can be described by the terms introduced in Chapter 2. For example, the dry bulb temperature *(T)* and absolute humidity *(H)* define many possible points, those air segments that are either saturated or partially saturated. For the purposes of this chapter, the use of the points above the saturation line is not permitted because the air is supersaturated and condensation must occur.

Other air characteristics are usually included on the psychrometric chart (Table 4.1). These include relative humidity (H_r) and volume of the humid mixture *(V)*. Many psychrometric charts will display other air characteristics, including wet bulb temperature *(T)*, dew point temperature (T_d), and absolute humidity in other units such as grains of moisture per pound of dry air. Appendix 4 contains psychrometric charts of use to students.

Only two of the air segment characteristics (from among T, T_w, H_r, H, H, and V) are required to define a point on the psychrometric chart. The characteristics (beyond two properties) may be calculated from the two chosen, and, for simplicity, some properties will be omitted from the psychrometric charts.

If two psychrometric properties of a humid air atmosphere are known, then all others can be determined from the psychrometric chart, or they can be obtained using a calculation utility.

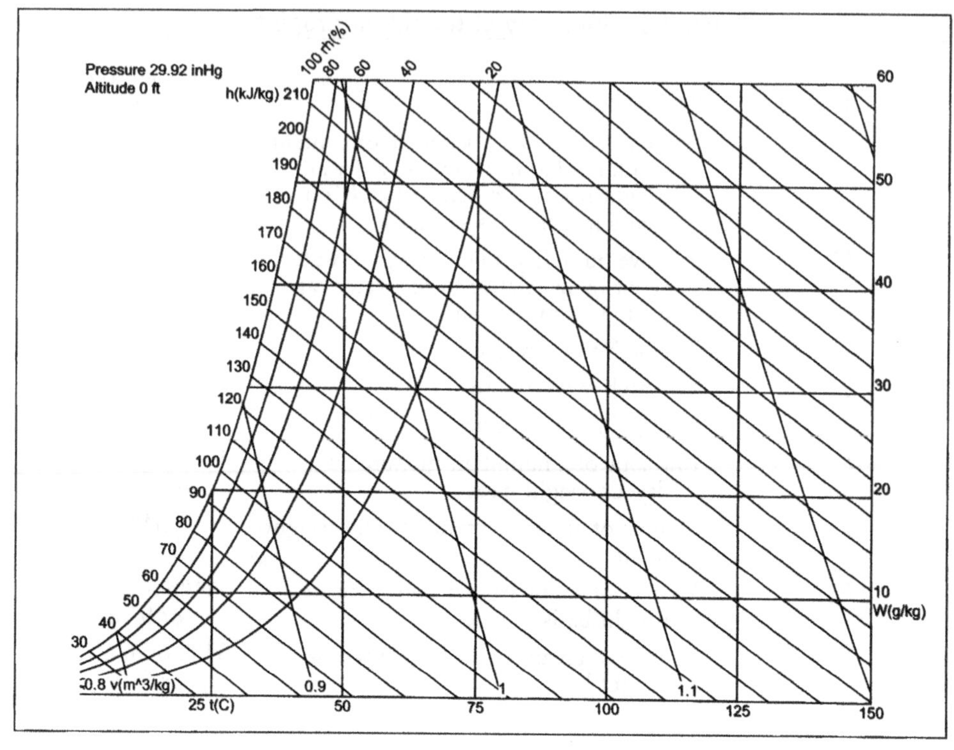

Figure 4.2: The psychrometric chart with SI units.

In this and subsequent chapters, the convention of a dry air basis for air properties is employed. For example, enthalpy and absolute humidity are expressed as Btu per pound of dry air and pounds of moisture per pound of dry air, respectively. In further simplification of these terms, Btu per pound of dry air will simply be represented as Btu/lb where it is understood that "lb" stands for pounds of dry air. The dimension "lb/lb" will similarly be understood to stand for pound of moisture per pound of dry air. In air conditioning problems, the term "grains of moisture" is used where 1 lb of moisture equals 7000 gr. The use of grains of moisture simplifies problems for air conditioning technicians because they do not have to deal with small or fractional numbers. In engineering practice, however, pounds of moisture or its metric equivalent of grams of moisture are used.

All psychrometric properties are displayed in English rather than metric (SI) dimensions in this chapter because they continue in predominant usage in North America. The properties may be conveniently converted, and some examples in this chapter are presented in both SI and English units for the convenience of the reader. The general form of the psychrometric chart is shown with SI dimensions in Fig. 4.2.

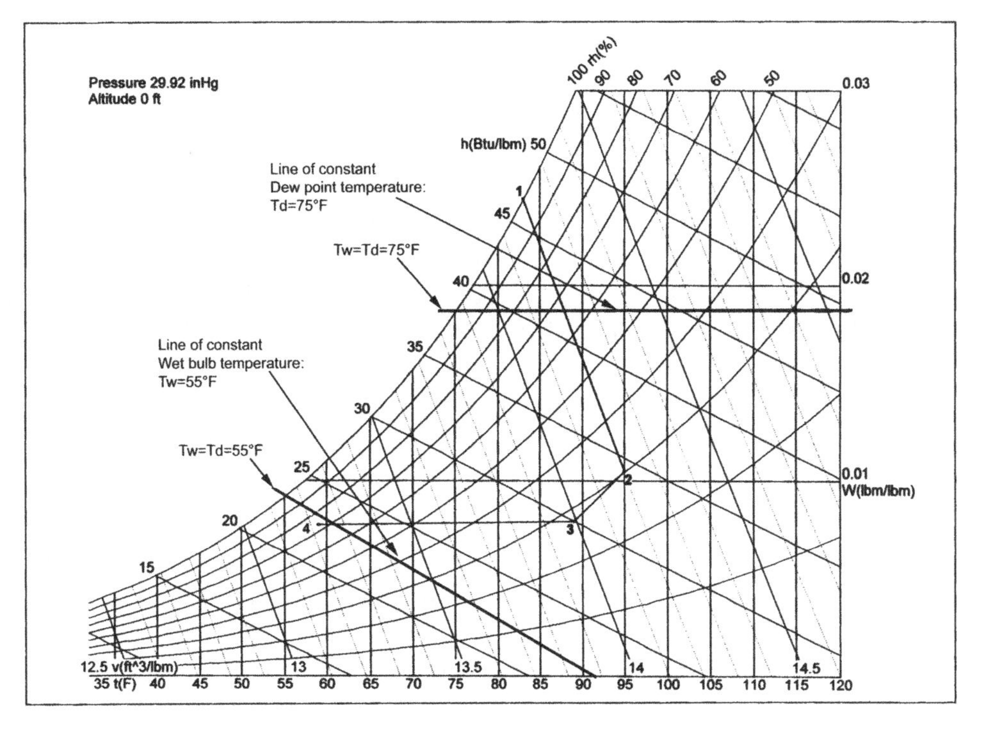

Figure 4.3: Locating points on the psychrometric chart.

Table 4.2: Properties of points in Fig. 4.3

Point	T	H_r	H	h	V	T_d	T_w
1	83	100	0.024	46.6	14.2	83	83
2	95	30	0.010	34.2	14.2	58.6	70.5
3	89	27	0.008	30	14	50.9	65.2
4	59	75	0.008	22.5	13.2	50.8	54.1

4.2 Basic Operations with Psychrometric Charts: Locating Points on the Psychrometric Chart

Each individual point on the psychrometric chart shown in Fig. 4.3 or on the saturation line can be used to represent and characterize an air segment that is involved in drying. Four points are located on Fig. 4.3 by the numbers 1, 2, 3, and 4. These points are connected with lines for convenience. Later on, connecting points on the chart will be employed to indicate a transition of an air segment involved in the drying process.

The properties of air segments displayed on this chart are T, H_r, H, h, and V. Any two of these are required to locate a point, and the others can be then read from the chart. Other properties not displayed may be determined by calculation or by reference to the traditional chart (Appendix 4). It is helpful to visualize a pyschrometric chart as a map where two points, such as longitude and latitude, are required to specify a location. In a similar manner, a psychrometric chart is a map of temperature and absolute humidity where two coordinates are required to specify a position. Examples of locating points and determining properties from Fig. 4.3 are discussed below.

Point 1: The point is located at T = 83°F and H_r = 100%. It is important to immediately note that the point is on the saturation line because the air segment is at 100% relative humidity. In addition, because the point is on the saturation line, T_d and T_w (not shown on this chart but calculated using psychrometric software) equal the dry bulb temperature *(T)* — that is, at saturation $T = T_w = T_d$.

At saturation, $T = T_w = T_d$

V and h are determined by interpolation because they are between the ruled lines indicating the values of these properties. One way to interpolate their approximate position is using a ruler and reading the smallest scale available (such as the mm scale). In the case of the volume property, a simple ratio can be set up in the following manner using the lowest (and nearest) volume line as the reference:

$$\frac{\text{Increment at Point 1}}{0.5\ \text{ft}^3/\text{lb}} = \frac{\text{Shortest distance from the 14 ft}^3/\text{lb line to Point 1}}{\text{Shortest distance from the 14 ft}^3/\text{lb line to 14.5 ft}^3/\text{lb line}}$$

where "shortest distance" refers to a line drawn perpendicular to the ruled (known) property lines. Then, the volume property is determined using the increment as follows:

$$V_{\text{Point 1}} = 14.0\ \text{ft}^3/\text{lb} + \text{Increment at Point 1}$$

Interpolation is used to locate points between ruled lines on psychrometric charts.

As will be shown in this chapter, various computer programs are available that provide the exact psychrometric properties of points on the chart. This has allowed the pyschrometric charts (Appendix 4) to be simplified to show only a few properties since the computer output provides the omitted data.

Point 2: The point is located at T = 95°F and H_r = 30%. Note that translation away from the saturation line reduced the relative humidity (H_r). There was a corresponding decrease in the dew point temperature (T_d) because the air now carries less moisture (as indicated by H). Now, the air also carries less energy, as indicated by the enthalpy *(h)*, even though the air segment of Point 2 has a higher dry bulb temperature *(T)* than that for Point 1.

It is interesting to compare the difference between the dry bulb and wet bulb temperatures for Points 1 and 2:

Point 1: $T - T_w = 83 - 83 = 0°F$ with $H_r = 100\%$

Point 2: $T - T_w = 95 - 70.5 = 24.5°F$ with $H_r = 30\%$

The difference $(T - T_w)$ is called the *wet bulb depression* (see Chapter 2, Section 2.13). As indicated earlier, relative humidity sensors have replaced wet bulb thermometers in many drying applications. Nevertheless, it is important to note that as the difference $(T - T_w)$ increases, the relative humidity decreases.

Point 3: The point is located at $T = 89°F$ and $H_r = 27\%$. Notice that because the distance between Points 2 and 3 is less than that between Points 1 and 2, the enthalpy *(h)* difference between Points 2 and 3 is also less than between Points 1 and 2. With experience, quick estimates of enthalpy changes can be made by simply looking at the chart — remembering that the magnitude of enthalpy changes should be judged based on the perpendicular distance from any given (ruled) enthalpy line chosen as an arbitrary reference.

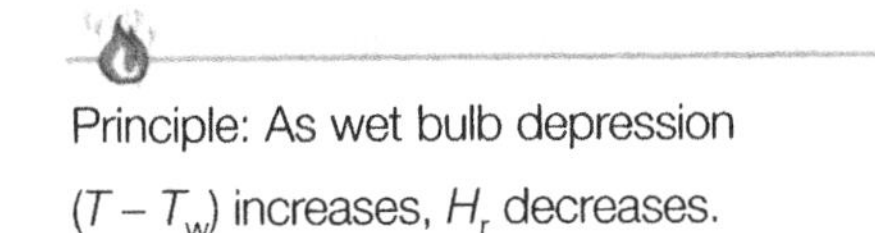

Principle: As wet bulb depression $(T - T_w)$ increases, H_r decreases.

Point 4: The point is located at $T = 59°F$ and $H_r = 75\%$. Notice that Points 3 and 4 exhibit the same absolute humidity of 0.008 lb/lb. If an air segment is changed from the conditions of Point 3 to those of Point 4, there is no loss or addition of water vapor from or to the air segment. This means that there is no latent energy change, since water vapor carries latent energy with it and there was no change in moisture content in this change. Therefore, the change from Point 3 to Point 4 involves only a sensible energy change, which is due to the dry bulb temperature difference.

Locating Wet Bulb and Dew Point Temperatures: Many older psychrometric charts provided a temperature scale on the saturation line where both T_w and T_d could be read directly on the chart. The computer-generated chart shown in Fig. 4.3 provides only a dry bulb temperature scale and an enthalpy scale along the saturation line because the computer directly provides both the wet bulb and dew point temperatures in a separate data box.

Lines of constant dry bulb temperature (isothermal lines) are shown as vertical lines on the chart. The wet bulb temperature, T_w, is the same as the dry bulb temperature only where T intersects the saturation line. In the case of a saturated atmosphere, the wet bulb temperature can be manually located by simply following the isothermal line from the saturation line down to the dry bulb temperature axis. Two

READING THE PSYCHROMETRIC CHART:

- Dry bulb temperature (T) is read along vertical lines from the scale on the bottom of the chart.
- Absolute humidity (H) is read along horizontal lines from the vertical scale on the right side of the chart.
- Enthalpy (h) is read along the scale provided at the saturation line.
- Wet bulb temperature (T_w) is established at the saturation line where the dry bulb temperature intersects. T_w is read along lines (not usually shown) extending from the saturation line and at 45° from the horizontal (parallel to isoenthalpic lines).
- Dew point temperature (T_d) is established at the saturation line where the dry bulb temperature intersects. T_d is read along lines (not usually shown) extending from the saturation line and parallel to the horizontal axis (parallel lines of constant H).

examples are provided in Fig. 4.3 (T_w = 55°F and T_w = 75°F). Once the wet bulb temperature is located, it can be read in any other analysis situation as lines of constant wet bulb temperature progressing from the saturation line and at 45° to the horizontal (parallel to the isoenthalpic lines). Lines of constant wet bulb temperature are typically not displayed on computer-generated charts.

The dew point temperature is read along horizontal lines from the saturation line across the chart. Once the dew point temperature is located on the saturation line, it can be read along a line of constant dew point temperature progressing from the saturation line and parallel to the horizontal axis (parallel to the lines of constant *H*).

Lines of constant enthalpy (isoenthalpic lines) are displayed at a 45° elevation from the baseline or *T* scale in Fig. 4.3. These isoenthalpic lines intersect the saturation line where the value of enthalpy can be read. The enthalpy is constant for any humid gas mixture along an isoenthaphic line.

Latent and Sensible Enthalpy Changes: Transitions of air segments on the psychrometric chart involve both latent and sensible energy changes depending on whether water is added or removed from the system and on temperature changes. A simple example related to drying of ceramics is shown in Fig. 4.4.

In this example, the dryer supply air (also called the hot supply) is maintained at 200°F at H_r = 0.4% (Point 1). The dryer exhaust air is found to exhibit T = 70°F and H_r = 70% (Point 2). Assume that no air leaks from the dryer and no energy is transferred through the dryer walls or doors (it's adiabatic).

TRANSITIONS:

- Horizontal transitions on the chart represent only sensible energy changes.
- Vertical transitions on the chart represent only latent energy changes.
- Transitions at some angle of inclination that is neither vertical nor horizontal involve both sensible and latent energy changes.

In order to further understand this process, a third point (Point 3) is located on the chart at the intersection of a horizontal line from Point 1 and a vertical line from Point 2, forming a right triangle with the three points. Because the energy change is the same for the

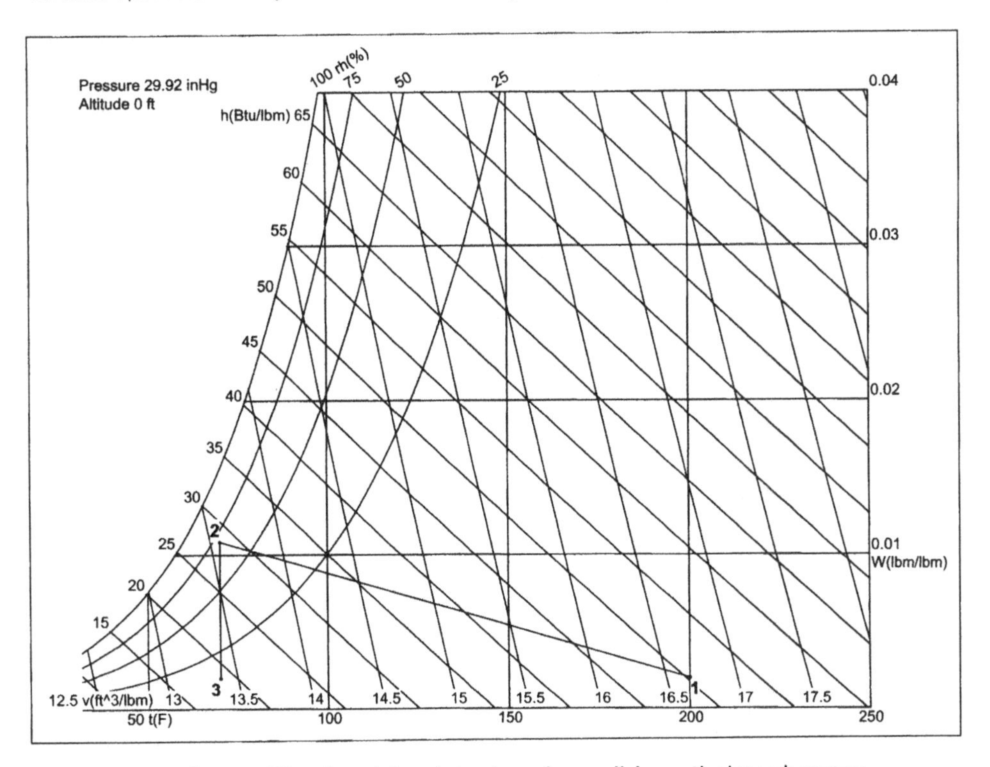

Figure 4.4: A transition involving latent and sensible enthalpy changes.

Table 4.3: Properties of points in Fig. 4.4

Point	T	H_r	H	h	V	T_d	T_w
1	200	0.4	0.019	50.3	16.7	18.7	86.1
2	70	70	0.002	28.6	13.6	59.8	63.3
3	70	13	0.002	18.9	13.4	18.7	47.7

transition Point 1 → Point 2 as it is for Point 1 → Point 3 → Point 2 for an adiabatic process, it is convenient to first consider the transition Point 1 → Point 3 (since it is only a sensible change). Then, it is convenient to consider the transition Point 3 → Point 2 (since it is only a latent change). The ideas in Table 4.3 are very important in analyzing transitions of air segments using psychrometric charts such this one. The property data for the points in Fig. 4.4 are given in Table 4.3.

The enthalpy difference between Points 1 and 2 is 21.7 Btu/lb. This same difference is computed by the sum of the enthalpy difference between Points 1 and 3 (31.4 Btu/lb) and Points 3 and 2 (−9.7 Btu/lb).

•••••

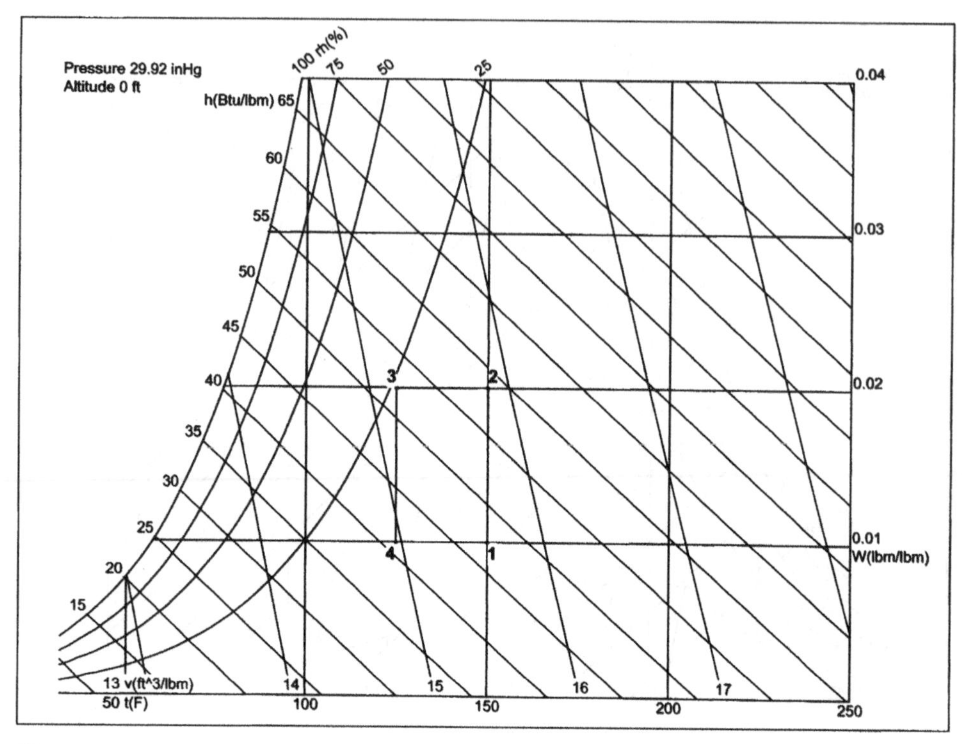

Figure 4.5: Example of air segment transitions using the psychrometric chart.

Basic operations of heating and cooling an air segment are further illustrated in Fig. 4.5. The individual operations are as follows:

Latent Heating of an Air Segment (Point 1 → Point 2): Point 1 on the psychrometric chart has the characteristics T = 150°F and H = 0.01 lb/lb. Other properties can be read on the chart: H_r = 6.3% and h = 47.2 Btu/lb. Properties that can be obtained through calculation are T_w = 83.5°F, T_d = 57.5°F, and V = 15.6 ft³/lb. Point 2 on the psychrometric chart has the characteristics T = 150°F and H = 0.02 lb/lb. Other properties can be read on the chart: H_r = 12.5% and h = 58.5 Btu/lb. Properties that can be obtained through calculation are T_w = 92.2°F, T_d = 77.1°F, and V = 15.85 ft³/lb.

The transition of the air segment from Point 1 to Point 2 involved humidification of the air segment, that is, adding water vapor to the air. In this particular transition, the air segment was maintained at a temperature *(T)* of 150°F. The addition of water vapor also added energy to the air segment as its enthalpy increased from 47.2 to 58.5 Btu/lb. Such transitions are typical in drying because drying air increases in moisture content. Enthalpy content of drying air may stay constant as sensible enthalpy is transferred into the product being dried in order to accomplish evaporation and latent enthalpy (in the form of water vapor) is transferred to the air. Note that the dew point, T_d, of the air segment increased because of the addition of moisture to the air segment.

It is important to note that the changes due to the transition of the air segment (Point 1 → Point 2) can be read directly from the psychrometric chart or can be obtained by calculation. Reading the changes directly from the chart or a calculation utility is much easier. The calculation is illustrated in Example 4.1.

EXAMPLE 4.1: Calculation of the enthalpy change in the segment (Point 1 → Point 2).

Given: Point 1 with H = 0.01 lb/lb, T_w = 84.3°F, and h = 47.2 Btu/lb.

Add 0.01 lb of water per pound of dry air without heating or cooling the air segment to reach Point 2. This involves an addition of latent heat to the air.

The latent heat (h_l) added that is due to the water vapor may be estimated using the relationship given below:

$$h_l \text{ (Btu/lb)} = 1092 - 0.55\ T_w$$

Therefore,

$$h_l = 1092 - 0.55\ (83.5) \sim 1049 \text{ Btu/lb}$$

and

$$(0.01 \text{ lb/lb})\ (1049 \text{ Btu/lb}) \sim 10.5 \text{ Btu/lb}$$

The exact enthalpy change, read from the chart, of 11.3 Btu/lb (58.5 – 47.2 Btu/lb) is comparable to the estimated enthalpy change of 10.5 Btu/lb obtained by calculation. The reason for the discrepancy with the calculation is the inexact estimation using the formula for the enthalpy change.

Sensible Cooling of an Air Segment (Point 2 → Point 3): The air segment of Point 2 is cooled to Point 3 while holding H constant (Fig. 4.3). Point 3 on the psychrometric chart has the characteristics T = 125°F and H = 0.02 lb/lb. Other properties that can be read on the chart are H_r = 23.8% and h = 52.3 Btu/lb. Properties that can be obtained through calculation are T_w = 87.6°F, T_d = 77.1°F, and v = 15.2 ft³/lb.

The transition of the air segment from Point 2 to Point 3 involved sensible cooling of the air segment; that is, there was no latent energy change because H was held constant. In this particular transition, the air segment was maintained at H = 0.02 lb/lb.

The cooling of the air decreased energy of the air segment as its enthalpy decreased from 58.5 to 52.3 Btu/lb, a net enthalpy decrease of 6.2 Btu/lb. Note that other properties of the air segment changed as well. For example, the air volume, V, decreased from 15.85 to 15.2 ft^3/lb and T_w changed from 92.2 to 87.6°F. Since no water was added or removed from the air segment, the dew point, T_d, remained constant at 77.1°F.

EXAMPLE 4.2: Calculation of enthalpy change in the segment (Point 2 → Point 3).

As in the case of the transition Point 1 → Point 2, an estimate of the sensible energy change in the transition Point 2 → Point 3 could be made by calculation rather than by reading the psychrometric chart or using a calculation utility. In brief, the specific heat of dry air is 0.24 cal/g-°C, or 0.77 Btu/lb-°F. The specific heat of water vapor is 0.46 cal/g-°C or 1.49 Btu/lb-°F. The problem is easily worked using 454 g of dry air (1 lb) and 9 g of water vapor (0.02 lb).

The energy change can be represented, as in Chapter 1, Section 1.6, as:

$$\begin{aligned}\text{Enthalpy change} &= \text{(mass of air) (specific heat)} (\Delta T) \\ &\quad + \text{(mass of water) (specific heat)} (\Delta T) \\ &= (454\text{ g}) (0.24\text{ cal/g-°C}) (13.9\text{°C}) \\ &\quad + (9\text{ g}) (0.48\text{ cal/g-°C}) (13.9\text{°C}) \\ &= 1572\text{ cal for 1 lb of dry air} \\ &\quad (6.24\text{ Btu for 1 lb of dry air})\end{aligned}$$

There is exact agreement between the enthalpy change read on the chart and the enthalpy change calculated from specific heat data. Note that the variation in specific heat with temperature was not considered in this calculation because it had a negligible effect. However, such corrections are appropriate for calculations involving higher temperatures. Tabulated data in enthalpy tables for air–water vapor mixtures could also be used in a calculation of such transitions.

Latent Cooling of an Air Segment (Point 3 → Point 4): The air segment of Point 3 is reduced in moisture content (dehumidified) in the transition to Point 4 while holding T constant (Fig. 4.3). Point 4 on the psychrometric chart has the characteristics T = 125°F and H = 0.01 lb/lb. Other properties that can be read on the chart are H_r = 12.1% and h = 41.1 Btu/lb. Properties that can be derived are T_w = 77.8°F, T_d = 57.4°F, and V = 15.0 ft^3/lb.

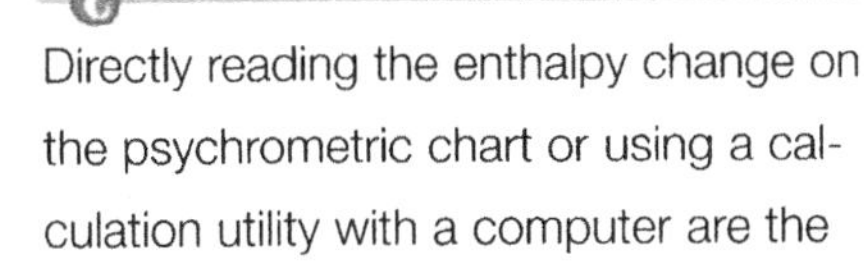

Directly reading the enthalpy change on the psychrometric chart or using a calculation utility with a computer are the easiest ways to obtain the difference in enthalpies in a transition of an air segment.

The transition of the air segment from Point 3 to Point 4 involves dehumidification of the air segment; that is, there was a latent energy decrease as the temperature *(T)* was held constant. In this particular transition, the air segment was maintained at T = 125°F. The dehumidification of the air decreased energy of the air segment as its enthalpy decreased from 52.3 to 41.1 Btu/lb, a net enthalpy decrease of 11.2 Btu/lb. Note that other properties of the air segment changed appropriately. For example, the air volume, T, decreased from 15.2 to 15.0 ft3/lb and T_w changed from 87.6 to 77.8°F. Since water was removed from the air segment, the dew point, T_d, decreased from 77.1 to 57.4°F

Completing the Loop by Sensible Heating of an Air Segment (Point 4 → Point 1): The return to Point 1 represents a sensible heating (no latent energy change) from Point 4 with 41.1 Btu/lb to Point 1 with 47.2 Btu/lb. The energy change is therefore (an addition) of 6.1 Btu/lb.

Summary of All Changes: All psychrometric properties and changes for the transitions from Point 1 to Point 4 are summarized in Tables 4.4 and 4.5.

The net energy change in the transition Point 1 → Point 2 → Point 3 →Point 4 → Point 1 is zero because the process was adiabatic (no energy changes outside of

Table 4.4: Psychrometric properties of air segments in Fig. 4.4

Point	T	H_r	H	h	V	T_d	T_w
1	0.01	150	6.3	15.6	47.2	57.5	83.4
2	0.02	150	12.5	15.8	58.5	77.1	92.2
3	0.02	125	23.8	15.2	52.3	77.1	87.6
4	0.01	125	12.1	15.0	41.1	57.5	77.8

Table 4.5: Enthalpy changes in transitions in Fig. 4.4

Transition	Process	Incremental change in enthalpy	Cumulative change in enthalpy
Point 1 → Point 2	Latent heating (humidification)	+11.27	+11.27
Point 2 → Point 3	Sensible cooling	−6.94	+5.03
Point 3 → Point 4	Latent cooling (dehumidification)	−11.17	−6.14
Point 4 → Point 1	Sensible heating	+6.14	0.00

A complete loop on a psychrometric chart results in a zero total enthalpy change. This is analogous to the Carnot cycle for an adiabatic process.

the system of the air segment). The enthalpy changes in Table 4.4 are further examined in Table 4.5. The incremental changes total 0.0 Btu/lb. In essence, this example has accomplished similar transitions as in the Carnot cycle (Chapter 1, Section 1.13).

4.3 Further Operations with Psychrometric Charts: Derived Units

As industrial fans are typically rated in cubic feet/minute (cfm), it is of interest to transform some properties to a cubic foot basis. For example, the data for Point 3 in Table 4.4 provides h = 52.3 Btu/lb, H = 0.02 lb/lb, and V = 15.2 ft³/lb (dry air basis). To change h and H to a per cubic foot basis, the following simple calculations are used:

$$\frac{52.3 \text{ Btu/ lb}}{15.2 \text{ ft}^3 \text{ / lb}} = 3.44 \text{ Btu/ ft}^3$$

$$\frac{0.02 \text{ lb moisture/ lb dry air}}{15.2 \text{ ft}^3 \text{ / lb}} = 0.0132 \text{ lb moisture/ ft}^3$$

Some computer programs for psychrometric analysis allow formation of special dimensions or quantities as needed.

4.4 Further Operations with Psychrometric Charts: Mixing Air Segments

Mixing air segments can be represented on the psychrometric chart by using the quantity parameters x and y where (as in Fig. 4.6):

x = the weight or volume of the component represented by a Point 1.

y = the weight or volume of the component represented by a Point 2.

L = the length of the line joining Points 1 and 2.

The weight (or volume) ratios (represented by capital letters) are then determined:

$$X = \frac{xL}{x + y}$$

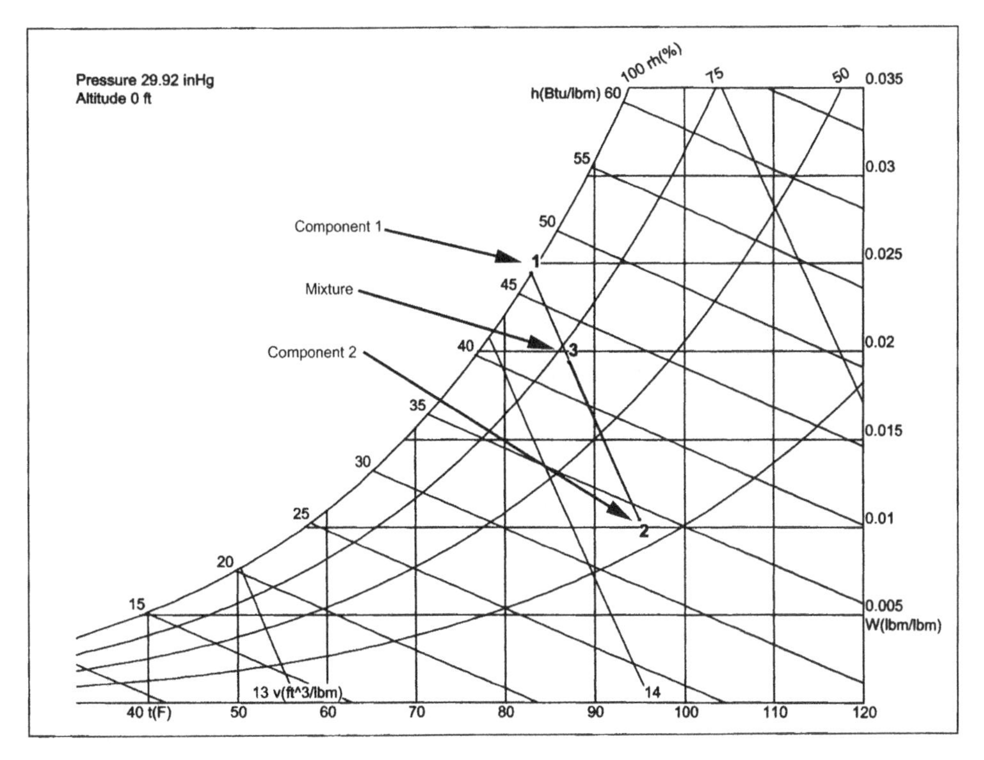

Figure 4.6: Example of mixing air segments.

Table 4.6: Properties of air segments in the mixing example (Fig. 4.6)

Point	T	H_r	h	V	H
1	83	100	46.65	14.2	0.0244
2	95	30	34.24	14.2	0.0104
3	87.2	70	42.13	14.2	0.0194

and

$$Y = \frac{yL}{x + y}$$

The resulting mixture is located on the line joining the air segments Points 1 and 2 in Fig. 4.6. The mixture must be located a distance X away from Point 2 and a distance Y away from Point 1. The resulting mixture is labeled Point 3 in Fig. 4.6. Property values for these points are given in Table 4.6.

Table 4.7: Example of psychrometric properties as a function of altitude (H=0.02 lb/lb)

Altitude (ft)	Altitude (m)	h	H_r	V	Btu/ft^3	Btu/ft^3 (% of 0 ft altitude)
0	0	59.54	2.05	16.86	3.531	100
1000	305	59.54	1.98	17.51	3.400	96
2000	610	59.54	1.91	18.16	3.279	93

4.5 Further Operations with Psychrometric Charts: Effect of Altitude

Altitude compensation for psychrometric quantities involving gas volumes may be important for higher elevations. Calculation utilities usually allow the user to set the altitude providing for accurate data. For example, properties of the same humid mixture with T = 200°F and H = 0.02 lb/lb are shown in Table 4.7.

4.6 Further Operations with Psychrometric Charts: Plotting Points on the Psychrometric Chart to Illustrate Changes

The process of understanding translations of air segments as their properties are changed — that is, moved across the psychrometric chart — is much easier if the points are physically plotted as in the above examples. This can be done by making photocopies of the charts and plotting the points or by using a computer program to plot the points.

Either physcially or electronically plotting points on the psychrometric chart is an important and essential technique in learning to use the charts in analyzing process changes.

4.7 Further Operations with Psychrometric Charts: Computer-Aided Interpretation of Psychrometric Charts

Computer software has become available that calculates the properties of air–water vapor mixtures. One such program is Humicalc™ (Thunder Scientific Corporation). The agreement between Humicalc information and that derived from conventional psychrometric charts is excellent at drying temperatures. Software suppliers typically comment on the accuracy of data at high temperatures generated using their programs.

Software that plots psychrometric charts and allows for insertion of points on the charts is provided by Akton Associates (Akton Psychrometric Chart™). With this software, properties of the air segment corresponding to the cursor position on the psychrometric chart are reported in a psychrometric properties window. This utility allows for instantaneous reading of the properties and also generates process reports.

PART B: ADVANCED CONCEPTS: SUMMARY OF EQUATIONS FOR PSYCHROMETRIC PROPERTIES

4.8 Introduction

The basis for the form of the psychrometric chart was discussed in Chapter 1, where it was stated that the saturation line is actually a phase boundary described by the Calusius-Clapeyron equation (Eq. 1.17):

$$\frac{dP}{dT} = \frac{\Delta h_v}{T \Delta V} \tag{4.1}$$

In order to transform the Calusius-Clapeyron equation into a more usable form, it is necessary to relate pressure (or partial pressure of water vapor, p_w) to absolute humidity *(H)*. This is done by recalling the definition of *H:*

$$H = \frac{m_w}{m_G} \tag{4.2}$$

where m_w and m_G are the weights of water vapor and dry gas in a humid air mixture. The total weight of the humid gas is then:

$$m_w + m_G = m_G(1 + H) \tag{4.3}$$

The ideal gas law can be stated, as below, for both the dry gas and water vapor:

$$m_G = \frac{p_G \mathfrak{v}}{RT} M_G \tag{4.4}$$

and

$$m_w = \frac{p_w \mathfrak{v}}{RT} M_w \tag{4.5}$$

where M_G and M_w are the molecular weights of the dry gas and water vapor, respectively. By substituting Eqs. 4.4 and 4.5 into Eq. 4.2, the result is:

$$H = \frac{p_w M_w}{p_G M_G} \tag{4.6}$$

Dalton's law of partial pressures states:

$$p = p_w + p_G \tag{4.7}$$

or

$$p_G = p - p_w \tag{4.8}$$

So by substituting Eq. 4.8 into Eq. 4.6, the result is:

$$H = \frac{p_w}{p - p_w}\left(\frac{M_w}{M_G}\right) \tag{4.9}$$

Because the molecular weights of water and air are 18 and 29 g/mol, respectively, Eq. 4.9 can be simplified as follows:

$$H = \frac{p_w}{p - p_w}\left(\frac{18}{29}\right) \tag{4.10}$$

Equation 4.10 allows the use of H in place of p (or p_w) in plotting the saturation line in the Calusius-Clapeyron equation. In other words, at any saturation water pressure (p_{ws}), there is a corresponding saturation absolute humidity (H_s). So the general form of the psychrometric chart using absolute humidity as the ordinate (vertical axis) is the same as when using pws as the vertical axis.

Equation 4.10 can be restated for saturation conditions as the definition of the saturation line or phase boundary:

$$H = \frac{p_{ws}}{p - p_{ws}}\left(\frac{M_w}{M_G}\right) \tag{4.11}$$

with the saturation boundary defined by the total pressure and the saturation vapor pressure of water at any given temperature, that is, the terms on the right side of Eq.

4.1. The relationship between p_{ws} and T for water vapor at 1 atm total pressure is also given in Chapter 2 (Table 2.5).

4.9 Percentage Humidity and Relative Humidity

The percentage humidity (H_p) is defined by the relationship:

$$H_p = 100\left(\frac{H}{H_s}\right) \tag{4.12}$$

By contrast, the percentage relative humidity (H_r) is defined by the relationship:

$$H_r = 100\left(\frac{p_w}{p_{ws}}\right) \tag{4.13}$$

4.10 Plotting Lines of Constant Relative Humidity on the Psychrometric Chart

Equations 4.13 and 4.10 can be combined to allow calculation of H at any H_r as follows:

$$H = \frac{H_r p_{ws}}{p - p_{ws}}\left(\frac{18}{29}\right) \tag{4.14}$$

4.11 Enthalpy of Water Vapor–Gas Mixtures

Since water vapor forms a mixture with dry gas or air (rather than a compound), the enthalpy can be simply stated as:

$$h_G = h_g + (h_r)(h_{ws}) + h_{mixing} \tag{4.15}$$

where: h_G = enthalpy of the humid gas mixture.
h_g = enthalpy of dry gas or dry air.
h_{ws} = enthalpy of water vapor at saturation.
h_{mixing} = enthalpy associated with the mixing of the water vapor and gas.

Since it is found that h_{mixing} is much less than h_g, the enthalpy of mixing is negligible, and Eq. 4.14 is restated as:

$$h_G = h_g + (H_r)(h_{ws}) \tag{4.16}$$

The enthalpy of the gas can be further developed considering how temperature changes affect the energy content of the gas mixture. To accomplish this, specific heat and enthalpy of vaporization will be employed in a gas segment heated from an initial temperature ($T_{initial}$) to a final temperature (T_{final}) considering a temperature of vaporization of the liquid (T_{vapor}). The enthalpy of the gas phase, h_G, then becomes:

$$h_G = C_{p,w}\left(T_{vapor} - T_{initial}\right) + \Delta H_v + C_{p,wv}\left(T_{final} - T_{vapor}\right) + C_{p,g}\left(T_{final} - T_{initial}\right) \tag{4.17}$$

where: $C_{p,w}$ = specific heat of water at constant pressure.
$C_{p,wv}$ = specific heat of water vapor at constant pressure.
$C_{p,g}$ = specific heat of dry gas (air) at constant pressure.

In words, Eq. 4.16 can be stated as: Enthalpy of the gas equals enthalpy increase of the water (first term in Eq. 4.17), plus the heat of vaporization added to change the water from a liquid to a gas phase (second term in Eq. 4.17), plus the enthalpy increase of the water vapor (third term in Eq. 4.17), plus the enthalpy increase of the air (fourth term in Eq. 4.17).

As the heat capacities are well known, it is a simple matter to calculate the enthalpy of the humid mixture given an assumption of $\Delta H_v = 0$ at the freezing point of water (0°C or 32°F). This latter assumption establishes a reference state, and the choice of the freezing point of water as the reference state was a matter of convenience.

Equation 4.17 can be restated in terms of saturation conditions for a humid air mixture:

$$h_G = C_{p,w}\left(T_{vapor,s} - T_{initial}\right) + \Delta H_v + C_{p,wv}\left(T_{final} - T_{vapor,s}\right) + C_{p,g}\left(T_{final} - T_{initial}\right) \tag{4.18}$$

where the subscript "s" indicates saturation conditions. But $T_{vapor,s}$ is the same as the dew point temperature (T_d), so Eq. 4.17 can be further simplified:

$$h_G = C_{p,w}\left(T_d - T_{initial}\right) + C_{p,wv}\left(T_{final} - T_d\right) + C_{p,g}\left(T_{final} - T_{initial}\right) \tag{4.19}$$

Equation 4.16 is restated for convenience:

$$h_G = h_g + (H_r)(h_{ws}) \tag{4.16}$$

and terms in Eq. 4.19 can be grouped so that it has the same form as Eq. 4.16 but generally applies to both saturated and unsaturated mixtures:

$$h_G = C_{p,g}\left(T_{final} - T_{initial}\right) + H_r\left[C_{p,w}\left(T_d - T_{initial}\right) + C_{p,wv}\left(T_{final} - T_d\right)\right] \quad (4.20)$$

Since $T_{initial} = 0$ (the reference point) and $T_{final} = T$ (the temperature of interest), Eq. 4.20 becomes:

$$h_G = C_{p,g}\left(T\right) + H_r\left[C_{p,w}\left(T_d\right) + C_{p,wv}\left(T - T_d\right)\right] \quad (4.21)$$

Equation 4.21 can be used to plot lines of constant H (isoenthalpic) lines on the psychrometric chart. The two key observations are that (a) enthalpy varies linearly with T and (b) enthalpy varies linearly with H_r.

4.12 Constructing Psychrometric Charts

Given the appropriate physical constants, psychrometric charts can be constructed using the key equations presented in this section and the ideal gas law. Computer-generated psychrometric charts use the same or similar equations, and these are usually provided in the user's manual included with the software.

4.13 Problems

In the following problems, the psychrometric properties include the following: T, H_r, H, h, T_d, T_w, and V. Assume 0 ft altitude (sea level) unless informed of a different altitude.

1. Complete the table below given two psychrometric properties in English units:

Point	T	Hr	h	Td*	Tw	V
1	80	20.0				
2	120		36.71			
3			57.56	74.1		
4	100				73.8	

**Requires a calculation utility*

2. Plot all four points from Problem 1 on a copy of the pyschrometric chart (copy Appendix 4 or generate a chart using a computer utility).

3. Assume you change properties of the air segments of Problem 1 in succession. Describe the changes in terms of the process of heating and cooling using the terms *latent* and *sensible:*

(a) Point 1 → Point 2

(b) Point 2 → Point 3

(c) Point 3 → Point 4

(d) Point 4 → Point 1

4. Given the transition shown below Point 1 (T = 75°F, H_r = 27.5%) → Point 2 (T = 140°F, H_r = 10%):

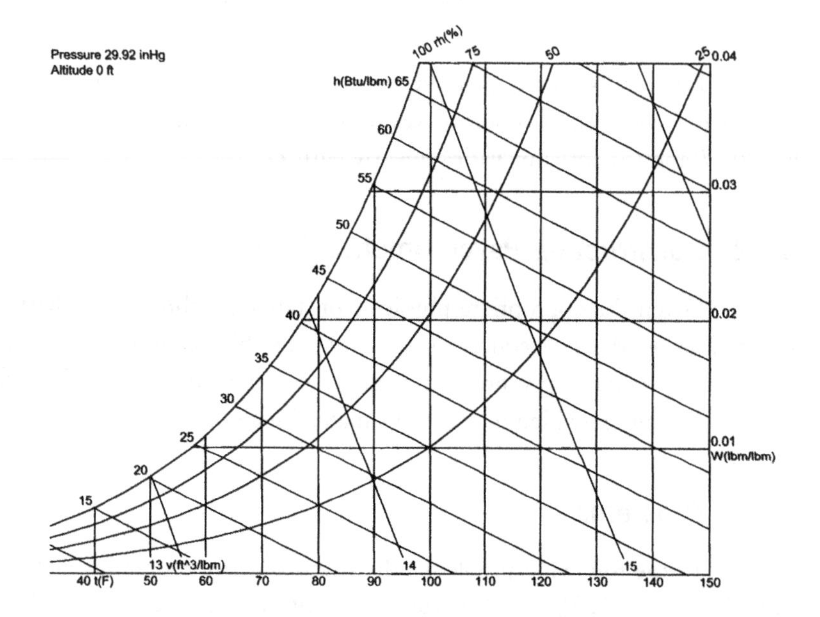

(a) Show on the diagram the sensible enthalpy change (h_s) (copy the diagram if necessary).

(b) What is h_s?

(c) Show on the diagram the latent enthalpy change (h_l).

(d) What is h_l?

5. What are h_s and h_l for the transition Point 2 → Point 4 in Problem 1?

6. What is the total enthalpy change in the transitions Point 1 → Point 2 → Point 3 → Point 4 → Point 1 in Problem 1?

7. Assume you mix 30 wt% of the air represented by Point 1 in Problem 1 with 70% of air represented by Point 3 in Problem 1. What are the psychrometric properties of the air mixture?

8. What are the psychrometric properties of the air segment represented by Point 3 of Problem 1 at an altitude of 1000 ft?

9. You are vacationing near the coast in the summer. When you wake up at 7:30 a.m., the temperature is 80°F, and you see condensation on the outside of the windows of your air-conditioned bedroom with the air conditioner thermostat set at 70°F (assume single glazing/poor thermal efficiency). The high at 1:00 p.m. is expected to be 97°F. If you want to play golf only when the grass is mostly dry, when should your tee time be?

10. You use a fan rated at 750 cfm (ft^3/min) to supply undercar air from a tunnel kiln to a predryer for special shapes (a small production line). If the undercar air exhibits $T = 130°F$ and $H_r = 25\%$, how many pounds of air per minute are you supplying to the predryer (on a dry air weight basis)?

11. A dryer exhaust stack condition is 80°F at a relative humidity of 90%. What happens to the product if, due to process changes, the exhaust temperature is reduced to 70°F given the same airflow and quantity of water evaporated during drying?

8. What are the [illegible] properties of the [illegible] segment [illegible] of [illegible] of 100 [illegible]

[illegible]

10. [illegible]

11. A [illegible] configuration [illegible] What [illegible]

CHAPTER 5

Characterization of Dryer Operations

PART A: ANALYZING DRYER PERFORMANCE USING PSYCHROMETRIC CHARTS

5.1 Introduction to Dryer Analysis

A dryer can be defined or described in a number of ways, but the most fundamental concept is that a dryer is a machine to accomplish energy transfer. In the usual case of a dryer processing ceramic products, the energy transfer is primarily by convection processes where sensible enthalpy (heat) is transferred to the product creating latent heat in the dryer atmosphere (as well as temperature increase in the product in Stage II of drying). In the case of microwave drying, energy in the form of radiation is transferred to the product, creating latent heat in the dryer atmosphere. Regardless of the source of energy, water vapor is added to the dryer atmosphere if drying is to be accomplished, thereby increasing the latent heat content of the atmosphere.

In considering dryers, the system of space must be expanded to include the dryer and its environment (as in Fig. 1.1, Choice A). In Chapter 4, the system was an air segment within a dryer, and changes within the air segment were adiabatic — no heat was lost to or gained from adjoining air segments. However, when the system is expanded to the dryer and its environment, the process is not adiabatic, as heat can be transferred to the environment surrounding the dryer.

Regardless of the method of applying energy to the product being dried, there must be air movement within a dryer to move the mass of evaporated water away from the product. This means that any dryer must have the following essential components:

- Hot supply: An entry point for air flowing to the dryer. In the case of a convection dryer, the hot supply obviously allows air at elevated temperature to enter the dryer, and that hot air is the energy source in drying. In the case of a dryer employing radiant energy, the supply air may or may not employ heated air. The convention in this chapter will be to call the supply the *hot supply* regardless of the type of dryer.

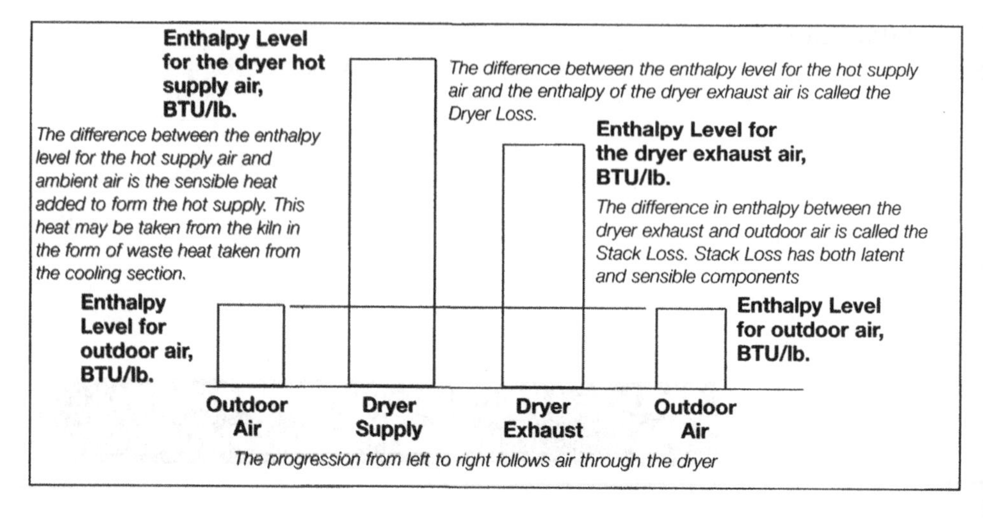

Figure 5.1: Representation of a dryer using a bar chart.

- Dryer exhaust: The exit point for air in the dryer to be taken or discharged to ambient in the dryer stack or smokestack. In real dryers air leakage occurs, and this form of air exiting the dryer will be included in *dryer loss* as defined in this chapter.

5.2 Representing Dryer Operation Using Bar Charts

The most convenient way to represent dryer performance is through bar charts that represent individual air segments in the dryer in terms of their enthalpy. In Fig. 5.1, the dryer operation is represented with ambient air heated to produce a hot supply condition. In the case of this dryer, the ambient air that is used is outdoor air. Both the outdoor air and the hot supply exhibit an enthalpy that can be determined by reference to the psychometric chart using at least two known properties of the air.

The Hot Supply: The air used for the hot supply may be called the *make-up air* since it makes up or creates the hot supply airflow. Enthalpy may be added by direct addition of heat through use of burners or resistive heating coils. In many cases, waste heat from the kiln is used as all or part of the make-up air for the dryer. In the case of waste heat, the ambient air is injected into the kiln cooling zone and removed prior to the hot zone to be carried through an insulated duct to the dryer hot supply. The conditions of the hot supply are a first major control point in the dryer.

The difference between the enthalpy of the hot supply air and outdoor (ambient) air is the heat added to the air to create the hot supply condition.

The Dryer Stack (Exhaust): Given a hot supplier condition, air passes through the dryer to the exhaust with an accompanying exchange of energy; that is, sensible enthalpy from the hot air is transferred to the product, causing evaporation and thus increasing the latent energy of the air. This air is primarily passed to the dryer exhaust for release or other use, and an exhaust condition is created. The properties of the exhaust air define the exhaust condition shown in Fig. 5.1.

The dryer loss is made up of all sources of energy lost from the dryer, not including the stack. It is obvious that a dryer will lose energy through its walls through conduction, through the insulation, and through external walls. Other important sources of energy loss include air leakage from the dryer. Every time an entrance or exit door is opened, air (carrying energy) escapes. Leakage from ducts can occur prior to air entering the dryer, through fans, or before establishing the stack condition. In addition, the dried product exiting the dryer, if exhibiting a temperature greater than the ambient condition, carries energy out of the dryer that could have been used in accomplishing drying. From the standpoint of this discussion, the energy carried out of the dryer by the dried product is considered a part of the dryer loss.

The difference between the enthalpy of the hot supply and the dryer exhaust (stack) is called the ***dryer loss.***

Stack Loss: Defining stack loss is useful to characterize dryer operation. It is necessary for a real dryer to exhibit stack loss — there must be a difference between the enthalpy of the dryer exhaust air and outdoor air, otherwise drying would not have taken place. The term *loss* can be confusing, and here *stack loss* simply refers to a necessary enthalpy difference between the stack and the ambient condition.

The difference in enthalpy between the dryer exhaust and outdoor air is called the ***stack loss.***

Stack loss includes both latent and sensible components. The latent stack loss is the difference in enthalpy between the stack and the outdoor condition due to the difference in water vapor content of the stack gases and the outdoor condition. Latent stack loss results from the process of drying — if no drying of the product occurred, there would be no latent stack loss. Sensible stack loss is due to the temperature difference between the stack gases and the outdoor air.

Simplifications in the Representation of Dryer Operation Using Bar Charts: The terms *ambient* and *outdoor air* have been used interchangeably in the above discussion. In many cases, the make-up air is supplied from within the plant. In winter operation, the plant air usually exhibits higher enthalpy than the ambient condition. In air-conditioned plants, the summer air can have lower enthalpy than the outdoor air. Appropriate adjustments in the representation of the dryer on the bar chart can be

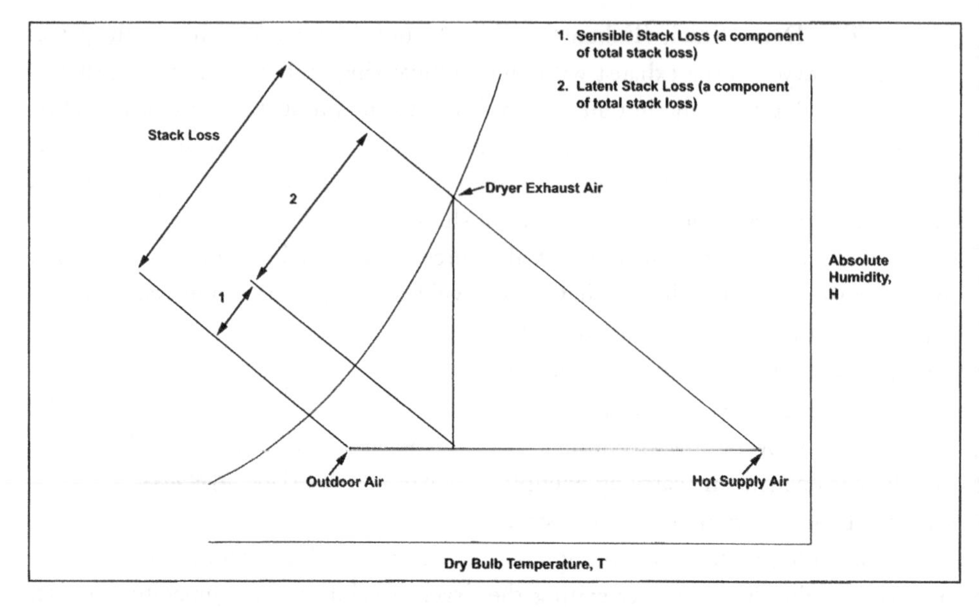

Figure 5.2: Dryer performance plotted on a psychrometric chart (100% air efficiency).

made. In any case, the definition of stack loss relative to outdoor air is the only important one relative to dryer performance.

5.3 Representing Dryer Operation on Psychrometric Charts

The bar chart representation of the dryer lacks only the data for the hot supply, exhaust, and ambient conditions. These are obviously input from instruments sensing these conditions, and the psychometric chart (or a psychometric software utility) is consulted to determine the enthalpies. Because the chart must be consulted to determine the utilities, it is interesting to plot the points that describe the dryer operation directly on the chart.

Case of a Dryer at 100% Air Efficiency of Air Utilization: The hot supply, exhaust, and ambient conditions for a dryer operating at 100% air efficiency (i.e., 100% efficiency in use of available air) are shown in Fig. 5.2. With respect to dryer operation, efficiency of air utilization is defined in terms of the extent of saturation of the air with water vapor as the air passes through the dryer. For a dryer to be operating at maximum (100%) efficiency, the air must be utilized to its maximum capacity; that is, it must be saturated with water vapor.

100% dryer efficiency occurs when exhaust air is saturated with water vapor.

After the supply, exhaust, and outdoor air points are plotted on the chart, the enthalpies can be read at the saturation line and can be added to the bar chart (Fig. 5.1). This is one very convenient method to visualize dryer operation. The same visualization can be seen in Fig. 5.2, where a line drawn from the exhaust enthalpy point to the outdoor air enthalpy point shows the stack loss.

100% dryer air efficiency occurs when exhaust air is saturated with water vapor.

In Fig. 5.2, the outdoor air and the hot supply air have the same absolute humidity *(H)*, and a horizontal line can be drawn connecting them, forming a baseline. This situation is duplicated when electric heating is used for the hot supply (no water vapor added to the make-up air) or when heated air is taken from the cooling zone of a tunnel kiln (with no reverse airflow or backdrafting).

A vertical line can be drawn from the exhaust condition to the baseline, allowing the sensible and latent components of the stack loss to be determined. This is similar to the process shown in Fig. 4.4, where a transition on the psychometric chart was broken into sensible and latent components.

The stack loss is the sum of its sensible and latent components. In the case of a dryer at 100% air efficiency, the latent stack loss shown is the maximum possible latent stack loss given the outdoor air conditions. The stack loss is at a maximum because no further water vapor can be added to the air passing through the dryer.

The maximum possible latent stack loss is observed when the dryer is at 100% air efficiency.

It will be shown that no real dryer can operate at 100% air efficiency. This is because ideal insulation does not exist and air leakage is always present. Nevertheless, the latent loss at 100% air efficiency is an important quantity for gauging the efficiency of a real dryer.

Latent loss at 100% air efficiency will be used to characterize a real dryer.

Heat/Energy Exchange of Air in Passage through a Dryer: It is useful to use an adaptation of Fig. 5.2 to illustrate energy exchanges of air during passage through the dryer. In Fig. 5.3, the total energy added to outdoor air to make up the hot supply is shown along the baseline between the outdoor air characteristics and the hot supply.

It is obvious that the air cools during the passage through dryer because heat is transferred into the product to accomplish evaporation. This means that the air in the dryer loses sensible heat. Because evaporation takes place, the air in the dryer gains latent heat. If the dryer is 100% efficient, the sensible loss for air in the dryer equals the latent gain for that air.

Dryer Operating at Less Than 100% Air Efficiency: As in the previous examples, the hot supply, exhaust, and ambient conditions for a dryer operating at less

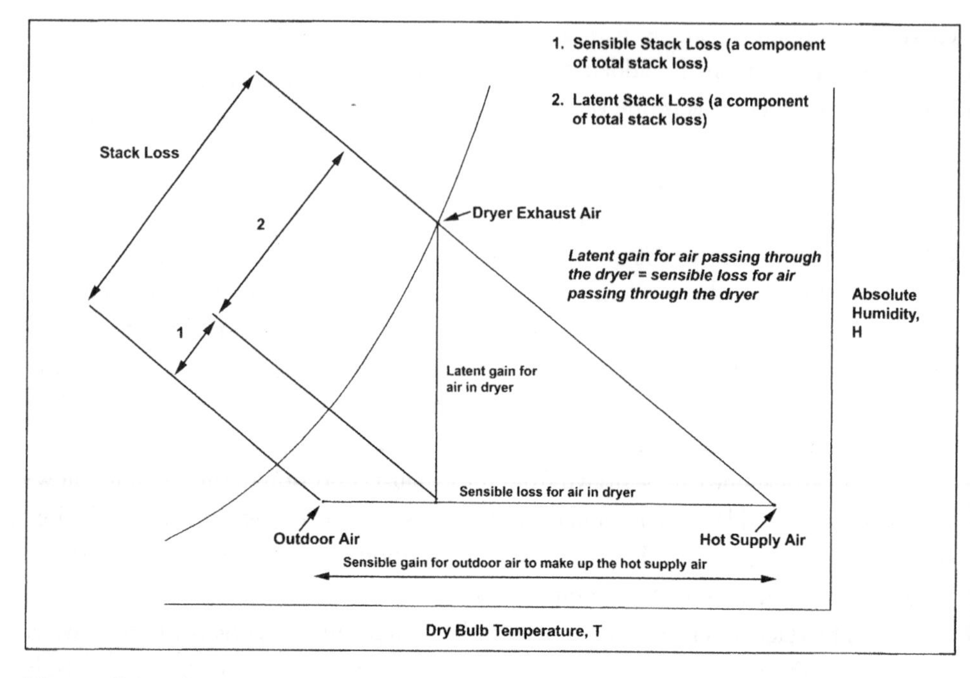

Figure 5.3: Energy exchange for air passing through a dryer at 100% air efficiency.

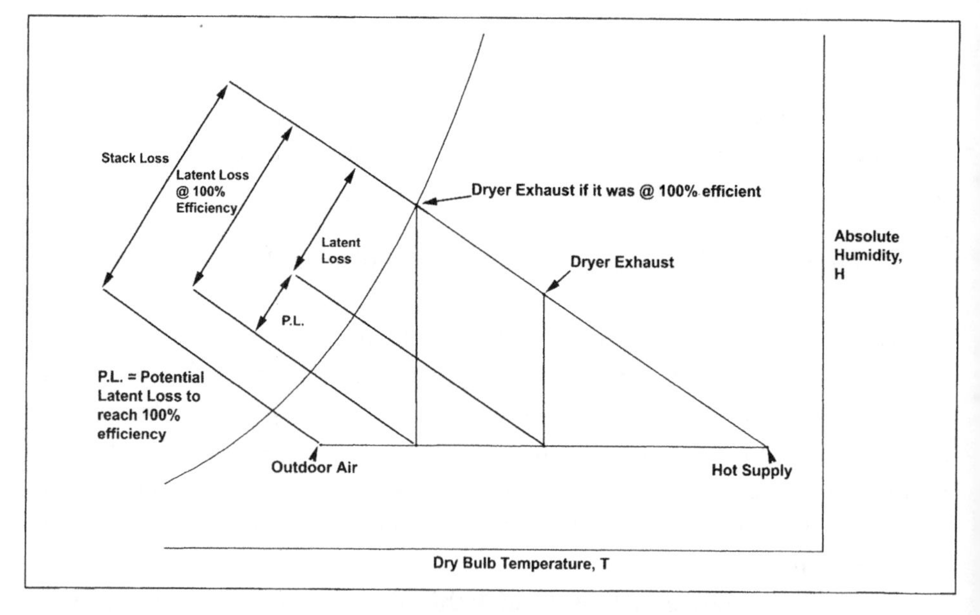

Figure 5.4: Dryer operating at less than 100% air efficiency.

than 100% air efficiency are shown in Fig. 5.4. The stack loss is determined by the usual procedure in reading the enthalpy difference between the dryer exhaust condition and the outdoor air.

If the dryer were operating at 100% air efficiency, the latent stack loss would be determined by projecting the dryer exhaust condition to the saturation line. Using this theoretical exhaust condition, the latent stack loss for this condition can be found by constructing a line down from the point to the baseline where the enthalpy can be read. This establishes what the latent stack loss would be if the dryer were operating at this condition.

The difference between the actual latent stack loss and the latent loss if the dryer were at 100% efficiency is called the *potential latent loss* (indicated by "PL" in Fig. 5.4). Once the potential latent loss is known, the dryer air efficiency can be calculated from the psychrometric data as follows:

$$\text{Dryer air efficiency} = \frac{\text{Actual latent stack loss of the dryer}}{\text{Latent stack loss if the dryer were at 100\% efficiency}} \times 100$$

Another method to calculate dryer efficiency (thermal efficiency) will be given in Part B. Regardless of the method chosen, the dryer efficiency calculation may be made from several perspectives. Dryer air efficiency basically indicates how air was used within the dryer in order to accomplish evaporation. With good air circulation, more evaporation will take place. If there is excessive air leakage, efficiency will drop.

DRYER AIR EFFICIENCY INDICATES:

- Air circulation efficiency within the dryer.
- Excessive air leakage or increasing air leakage.
- The need for seasonal dryer adjustments.

EXAMPLE 5.1:

Given a continuous countercurrent dryer with the following operating parameters:

Ambient air: 65°F dry bulb at 60% relative humidity (H_r) (T_w = 55°F)

Hot supply air: 330°F dry bulb at 0.9% relative humidity (T_w = 110°F)

Dryer exhaust: 120°F dry bulb at 34.4% relative humidity (T_w = 90°F)

Questions:

1. What is the dryer loss?

2. What are the latent, sensible, and total stack losses for this dryer?

3. What would the stack loss be for this dryer if it were at 100% efficiency?

4. What is the actual dryer air efficiency?

5. How much air (or sensible heat) was available for drying but not used in this dryer?

6. How much air (or sensible heat) was not available for use in drying in this dryer?

Solution Techniques:

1. Plot the dryer operating characteristics on the psychometric chart (Fig. 5.5).

2. Identify the enthalpy at each key operating point, and construct and label a bar chart for this dryer.

3. Calculate the dryer air efficiency.

4. Determine the sensible heat available but not used (potential latent loss).

5. Calculate the sensible heat not available for drying (sensible loss at 100% efficiency).

To complete the problem, it is convenient to assemble the bar chart for it.

Solutions (points on Figs. 5.5 and 5.6):

1. What is the dryer loss?

  Given by Point 2 – Point 3:
  88.17 Btu/lb – 55.42 Btu/lb = 32.75 Btu/lb

2. What is the latent, sensible, and total stack loss for this dryer?

  Total stack loss given by Point 3 – Point 1:
  55.42 Btu/lb – 23.15 Btu/lb = 32.27 Btu/lb

  Sensible stack loss given by Point 5 – Point 1:
  36.52 Btu/lb – 23.33 Btu/lb = 13.19 Btu/lb

  Latent stack loss given by Point 3 – Point 5:
  55.42 Btu/lb – 36.52 Btu/lb = 18.90 Btu/lb

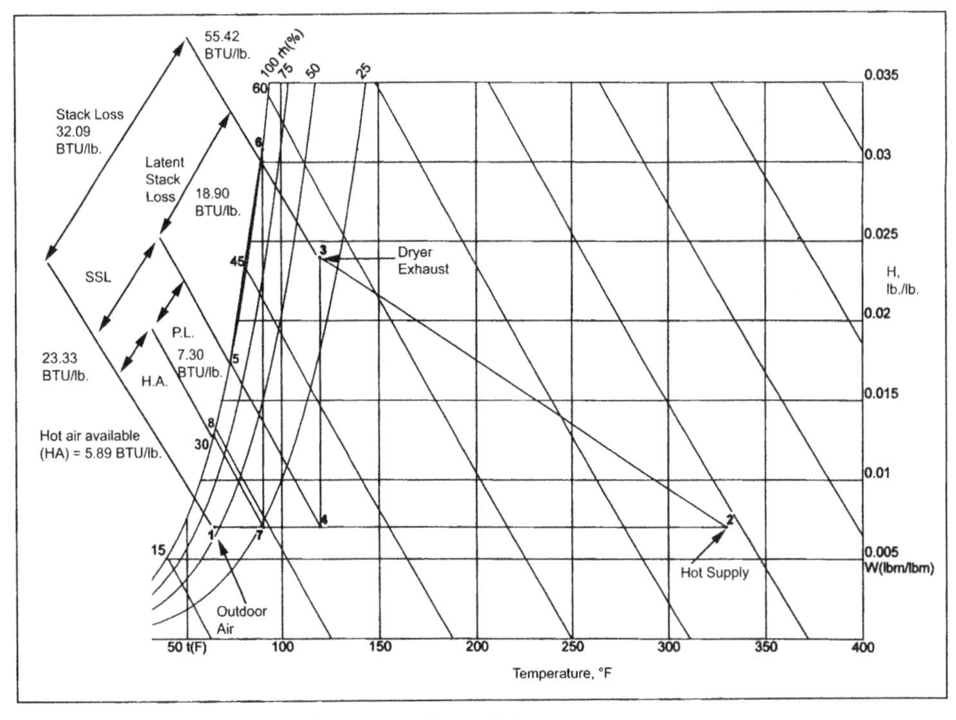

Figure 5.5: Solution to the example problem.

Table 5.1: Points on Fig. 5.5 (from Akton Psychometric Chart)

Point	Description	H (lb/lb)	T (°F)	H_r (%)	V (ft³/lb)	h (Btu/lb)	T_w (°F)
1	Outdoor air	0.007	65	53.8	13.36	23.33	55.1
2	Hot supply	0.007	330	0.16	20.11	88.17	109
3	Dryer exhaust	0.024	120	32.5	15.16	55.42	90
4	Projection	0.007	120	9.76	14.76	36.52	73.1
5	Projection	0.0174	73.1	100	13.79	36.52	73.1
6	Projection	0.0308	90	100	14.53	55.42	90
7	Projection	0.007	90	23.6	14	29.22	64.2
8	Projection	0.0128	64.2	100	13.46	29.22	64.2

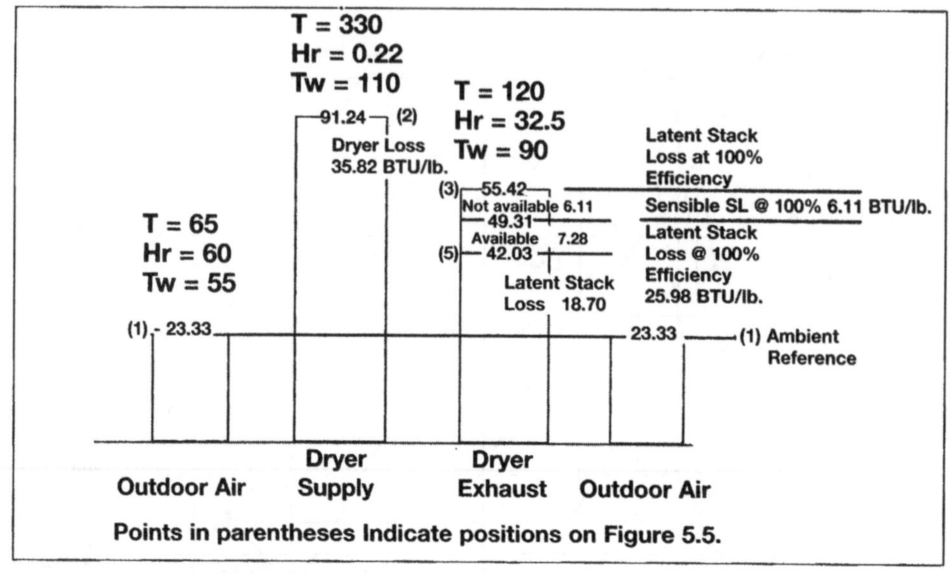

Figure 5.6: Bar chart for the example problem. (See data in Table 5.1. Slight differences between the table and the information here is due to use of different calculation programs.)

3. What would the stack loss for this dryer be if it were operating at 100% efficiency?

Given by Point 6 – Point 8:

55.42 Btu/lb – 29.22 Btu/lb = 26.20 Btu/lb

4. What is the actual dryer air efficiency?

$$\frac{\text{Actual latent stack loss}}{\text{Latent stack loss at 100\%}} \times 100 = \frac{13.37 \text{ Btu/lb}}{26.20 \text{ Btu/lb}} \times 100 = 51.0\%$$

5. How much air (or sensible heat) was available for drying but not used in this dryer?

Given by Point 5 – Point 8:

36.52 Btu/lb – 29.22 Btu/lb = 7.30 Btu/lb

6. How much air (or sensible heat) was not available for use in drying in this dryer?

Given by Point 8 – Point 1:

29.22 Btu/lb – 23.33 Btu/lb = 5.89 Btu/lb

or alternately

Sensible stack loss – available air = 13.19 Btu/lb – 7.30 Btu/lb

= 5.89 Btu/lb

PART B: SANKEY DIAGRAMS AND DRYER THERMAL EFFICIENCY

5.4 Representing Operating Dryers: Sankey Diagrams

Sankey diagrams are representations of thermal processes showing mass and heat flow. A diagram for a clay brick plant is shown in Fig. 5.7. This diagram shows a continuous countercurrent convection dryer using waste heat from the tunnel kiln where the dryer is attached to or a part of the kiln. This combination is called an *inline dryer/kiln.*

The kiln and dryer are separated by an internal door, which is raised whenever product is advanced (pushed). The inputs, outputs, and other movements of air are shown by use of arrows. Note that air mass and enthalpy are given per pound of fired product (ware) exiting the kiln. This is a convenient representation when the cost per pound of the fired product is of primary interest.

Some features of the Sankey diagram shown in Fig. 5.7 relative to the dryer operation are shown in Table 5.2. The energy used in drying can be calculated as shown in Table 5.3; energy losses are shown in Table 5.4.

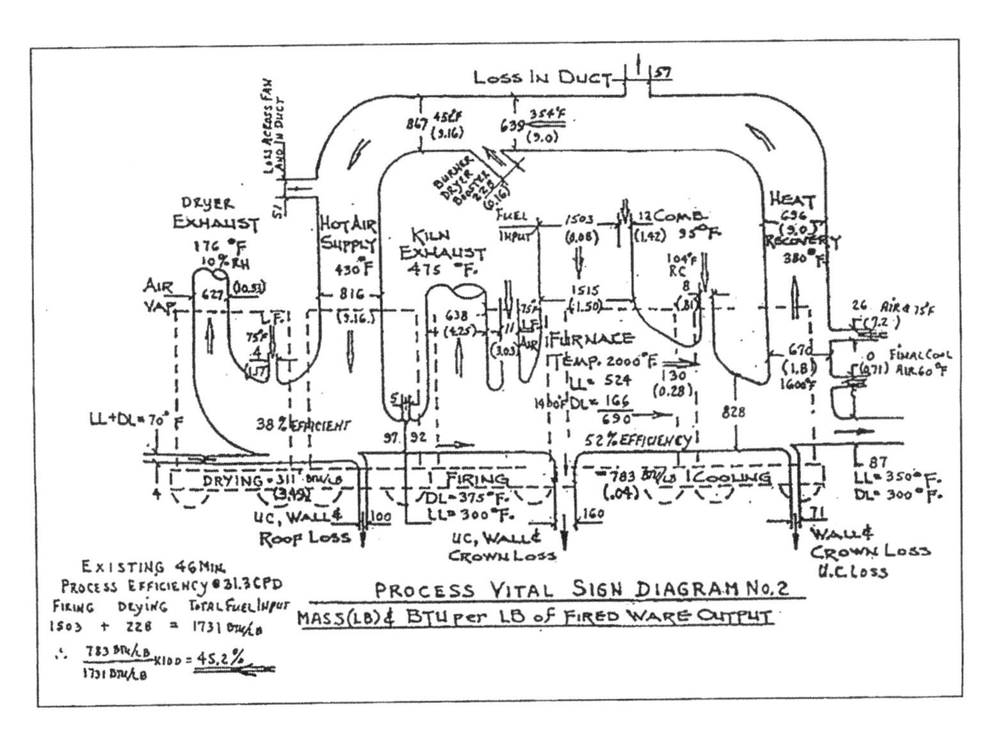

Figure 5.7: Sankey diagram for an operating dryer.

Table 5.2: Summary of data on the Sankey diagram in Fig. 5.7

Location	T (°F)	H_r	Btu/lb of fired product	Weight (lb) per lb of fired product
Hot supply	430		816	9.16
Dryer exhaust	176	10	627	10.53
Exiting product	NA		97	
Under car (UC), wall, and roof loss			100	3.49
Leakage factor (LF) input (fans, etc.)	75		14	
Leakage loss (LL) and door loss (DL)	70		4	
Loss at internal door			5	

Table 5.3: Energy used in drying (Fig. 5.7)

Hot supply input	+816 Btu/lb
Leakage input (LF)	+14 Btu/lb
Energy carried by product on exit	−92 Btu/lb
Energy output at dryer exhaust	−427 Btu/lb
Net energy used	311 Btu/lb of fired product

Table 5.4: Energy losses in drying (Fig. 5.7)

LL + DL	4 Btu/lb
UC, wall, and roof	100 Btu/lb
Internal door	5 Btu/lb
Total dryer losses	109 Btu/lb of fired product

The dryer operation represented in Fig. 5.8 can be analyzed using a psychometric chart given the original data and additional data on the relative humidity of outdoor air:

Outdoor air: $T = 75°F$ with $H_r = 30\%$ ($H = 0.0057$ lb/lb)
Hot supply: $T = 430°F$ with $H = 0.0057$ lb/lb
Dryer exhaust: $T = 176°F$ with $H_r = 10\%$

5.5 Dryer Thermal Efficiency

Thermal efficiency of the dryer discussed in Section 5.4 is calculated using the relationship:

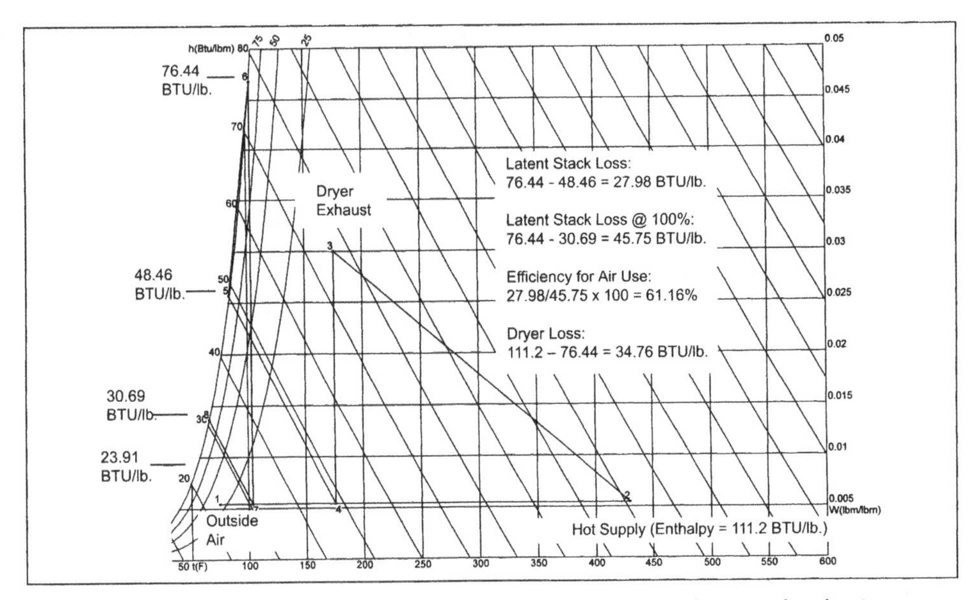

Figure 5.8: Analysis of the operating dryer using a psychrometric chart.

Table 5.5: Points on Fig. 5.8 (from Akton Psychrometric Chart)

Point	Description	H (lb/lb)	T (°F)	H_r (%)	V (ft³/lb)	h (Btu/lb)	T_w (°F)
1	Outdoor air	0.00547	75	30	13.58	23.91	56.4
2	Hot supply	0.00547	430	0.037	22.61	111.2	22.6
3	Dryer exhaust	0.03	176	10	16.78	76.44	103
4	Projection	0.00547	176	1.89	16.15	48.46	84.5
5	Projection	0.0257	84.5	100	14.27	48.46	84.5
6	Projection	0.00468	103	100	15.24	76.44	103
7	Projection	0.00547	103	12.4	14.29	30.69	66.1
8	Projection	0.00137	66.1	100	13.53	30.69	66.1

$$\text{Thermal efficiency (\%)} = \frac{\begin{array}{c}\text{Enthalpy calculated to dry product}\\ \text{(Btu / lb of dried product)}\end{array}}{\begin{array}{c}\text{Enthalpy consumed in drying product}\\ \text{(Btu / lb of dried product)}\end{array}} \times 100$$

As an example, the energy necessary to dry a clay ceramic product with a dried weight of 1 lb (454 g) and containing 12% moisture (DB) in the green state is shown below: (using data from Table 5.2):

Table 5.6: Efficiency criteria for dryers

Efficiency criteria	Formula	Efficiency value (%)
Thermal efficiency	$\frac{\text{theoretical heat requirement}}{\text{actual heat requirement}} \times 100$	54.1
Adjusted thermal efficiency (energy to accomplish drying)	$\frac{\text{theoretical heat requirement}}{\text{actual requirement} - \text{dryer losses}} \times 100$	73.0
Air efficiency	$\frac{\text{latent stack loss (Btu/lb. of air)}}{\text{latent stack loss @ 100\% efficiency}} \times 100$	61.6

- Initial temperature: 60°F (16°C); final temperature: 430°F (221°C).
- Water weight in product = (0.12) (454 g) = 36 g.
- Heat capacity of clay (0.22 cal/g-°C), water (1 cal/g-°C), and water vapor (0.46 cal/g-°C).
- Enthalpy of vaporization of water (540.5 cal/g).

$$\Delta H = \text{enthalpy to heat clay} + \text{enthalpy to heat water} + \text{enthalpy to evaporate water} + \text{enthalpy to heat water vapor}$$

$$\Delta H = (454\text{ g})(0.22\text{ cal/g-°C})(221\text{°C} - 16\text{°C}) + (36\text{ g})(1\text{ cal/g-°C})(100\text{°C} - 16\text{°C}) + (36\text{ g})(540.5\text{ cal/g}) + (36\text{ g})(0.46\text{ cal/g-°C})(221\text{°C} - 100\text{°C})$$

$$\Delta H = 20{,}475\text{ cal} + 4{,}536\text{ cal} + 29{,}187\text{ cal} + 3{,}005\text{ cal}$$

$$\Delta H = 57{,}203\text{ cal or } 227.1\text{ Btu for 1 lb of dried clay}$$

The value of ΔH, the enthalpy required to accomplish the drying in this example, can be viewed as the theoretical enthalpy to dry 1 lb of product. Because the actual process required 420 Btu/lb (311 Btu/lb plus the thermal losses from the dryer of 109 Btu/lb), the thermal efficiency can be calculated:

$$\text{Thermal efficiency (\%)} = \frac{227.1\text{ Btu/lb of dried product}}{420\text{ Btu/lb of dried product}} \times 100 = 54.1\%$$

To compare the thermal and air efficiencies, it is necessary to point out that the air efficiency calculation did not consider the dryer losses. The adjusted thermal efficiency, excluding dryer losses, becomes 227.1 Btu/lb divided by 311 Btu/lb, or

73.0%. The various efficiencies calculated for this dryer example are given in Table 5.6. It is obvious that the various efficiencies represent different things, and the choice of efficiency criteria may depend on the way the information will be used. Thermal efficiency may be more useful in comparing one type of dryer to another. Adjusted thermal efficiency may be more useful in comparing one type of material to another. Air efficiency may be more useful in comparing air circulation within a dryer, seasonal variations, or process drift (as in effect of air leaks).

5.6 Problems

All questions are based on 1 lb of dry air.

1. Calculation of heats in dryer. Given:

Ambient air:	T = 35°F; T_w = 35°F
Hot supply air:	T = 104°F; T_w = 65°F
Exhaust (stack):	T = 65°F; T_w = 65°F

   (a) How much sensible heat was added to ambient air (ambient to supply)?

   (b) How much latent heat was added to ambient air (ambient to supply)?

   (c) How much sensible heat was exchanged for latent heat in the dryer?

   (d) How much water was extracted from the drying product by the air?

   (e) What is sensible heat stack loss?

   (f) What is latent heat stack loss?

   (g) What is dew point temperature of dryer exhaust?

   (h) What is dryer loss (heat loss from air during passage through the dryer)?

2. Influence of change in exhaust conditions. Given: All conditions are the same as in Problem 1 except that the dryer exhaust changed to T = 75°F; T_w = 65°F.

   (a) What is the sensible heat stack loss?

   (b) What is latent heat stack loss?

   (c) How much water was picked up from the product?

   (d) What is dew point temperature of exhaust?

   (e) What is dryer loss?

3. Reduction in exhaust temperature. Given: All conditions are the same as Problem 1 except that the dryer exhaust changed to $T = 60°F$; $T_w = 60°F$.

(a) What is latent heat stack loss?

(b) What is sensible heat stack loss?

(c) What is quantity of water pickup?

(d) What is dryer loss?

4. Summertime operation. Given:

Ambient air	$T = 100°F$; $T_w = 95°F$
Hot supply	$T = 205°F$; $T_w = 110°F$
Water pickup	0.022 lb water per 1 lb air
Dryer loss	0

(a) What is the exhaust T_w?

(b) What is the exhaust T?

(c) What is the amount of heat added to ambient air?

(d) What flow is required (cfm) to supply 100 lb/min at the hot supply?

5. Wintertime operation. Given:

Ambient air	$T = 35°F$; $T_w = 31°F$
Hot supply	$T = 205°F$
Water pickup	0.022 lb water per 1 lb air
Dryer loss	0

(a) What is the T_w of the hot supply?

(b) What is T of the exhaust?

(c) What is T_w of the exhaust?

(d) How much heat is added to ambient air?

(e) What is the percentage sensible heat increase over Problem 4 (per lb of air)?

(f) How many pounds of air per minute would be supplied at the hot end by the 1760 cfm fan of Problem 4?

(g) What is the percentage increase in water removed per cubic foot in Problem 5 over Problem 4?

(h) What is the percent relative humidity in the dryer exhaust for Problems 4 and 5?

6. Wintertime operation with no change in available heat.

Ambient air	T = 45°F; T_w = 31°F
Production rate	0.022 lb water per min
Hot supply	1 lb air per min; same heat added as in Problem 4 per lb of air

(a) What is the T of hot supply?

(b) What is the T_w of hot supply?

(c) What is the T of the exhaust?

(d) What is the T_w of the exhaust?

(e) How much water will each pound of air pick-up in the dryer?

(f) What happens to the rest of the input water?

(g) What happens when outside air enters the charging end of the dryer?

(h) How much percentage increase in mass of air would be required to remove 0.022 lb of water/min?

7. Correcting for winter conditions by reducing production rate. Given:

Ambient air	T = 35°F; T_w = 31°F
Hot supply	1 lb air per min; same heat added as in Problem 4 per lb of air
Production rate	0.015 lb of water/min

(a) What is the percentage reduction in production rate from Problem 6?

(b) What is exhaust T?

(c) What is exhaust T_w?

8. Correcting for condensation by increasing quantity of air to keep the same relative humidity as in the exhaust as in problem 5 (45%). Given:

Ambient air	T = 35°F; T_w = 31°F
Exhaust air	T = 92°F; T_w = 76°F
Production rate	0.022 lb water/min
Hot supply air	T = 140°F; T_w = 76°F

(a) How many pounds of air per minute to remove 0.022 lb water per minute?

(b) What is the percentage increase in air over Problem 5?

(c) What quantity of heat is added to ambient air per minute?

(d) What quantity of heat was added to ambient air per 0.022 lb water in Problem 5?

9. Increasing production rate in summer operation. Given:

Production rate	Increase from 0.022 to 0.029 lb/min (32% increase)
Hot supply air	1 lb/min; $T = 205°F$, $T_w = 110°F$
Ambient	$T = 100°F$; $T_w = 950°F$
Dryer loss	0

(a) What is the exhaust T?

(b) What is the exhaust T_w?

(c) What is the water pickup in dryer per minute?

(d) What happens to excess water?

10. Decreasing production rate in summer operation. Given:

Production rate	Decrease from 0.022 to 0.015 lb/min
Hot supply air	1 lb/min; $T = 205°F$, $T_w = 110°F$
Ambient	$T = 100°F$; $T_w = 95°F$
Dryer loss	0

(a) What is the exhaust T?

(b) What is the exhaust T_w?

(c) What is the water pickup in dryer?

(d) What is the sensible heat stack loss?

(e) What is the percentage increase in sensible heat for Problem 10 over Problem 4?

(f) What is the sensible heat loss in 1000 brick (containing 800 lb water) for Problem 10?

(g) What is the sensible heat loss in 1000 brick (containing 800 lb water) for Problem 4?

(h) How many cubic feet of gas (1000 Btu/ft^3) would you save per thousand brick by changing operation from Problem 10 to Problem 4?

CHAPTER 6

Dryer Control: Controlling Countercurrent Convection Dryers

6.1 Introduction

Building brick and refractory manufacturers dry much of their production in car tunnel dryers. The heated air traveling countercurrent to the direction of car travel provides heat and air for water removal, and the thermal energy is supplied into the product primarily by convection. Control of the drying process can be challenging with this type of equipment.

The challenge, and even the need for control, will vary widely for different products and different manufacturing procedures. Some raw materials are easy to dry and can be inserted into an "abusive" fast drying environment. Other materials are very sensitive to the drying environment and require a closely regulated drying procedure to avoid cracking, loss of strength, warping or other deterioration in properties. Even some so-called easy-to-dry materials may produce internal flaws that reduce product performance, but because the cracks are not visible on the surface, the manufacturer believes it is a quality product.

The physical characteristics of the material being dried establish the tolerances of the drying process that in turn produce requirements for the level of control employed.

Drying consumes a large amount of thermal energy. Theoretically, it takes about 1000 Btu (1055 kJ) to evaporate 1 lb of water. Efficient brick dryers should use 1600–1800 Btu per pound of water evaporated, and thermal efficiency of any dryer using over 2000 Btu per pound of water can be improved. The addition of dryer controls can work not only to ensure the quality of the product, but also to improve the thermal efficiency of the equipment.

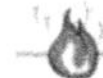

The best countercurrent convection dryers exhibit a thermal efficiency of about 55% to about 62%.

The cost and complexity of the control system increases with the number and extent of variables in the process. No control is required for a dryer with a constant rate of loading, a constant supply of heated air, and unchanging weather.

Sophisticated control is needed to accommodate changing production rates, changing product size, changing raw material composition, and changing weather.

6.2 The Single, Closed Tunnel Concept

The ceramic industry has a long history of using dryers that are a closed tunnel where heated air is introduced at one end and is exhausted, with its burden of water, at the opposite end of the dryer. The product travels in a direction countercurrent to the direction of air flow. In the past, there was little concern about the exit or entrance of air along the tunnel length. Some improvement of operation came from adding fans to move air from the top of the setting in one zone and reintroduce it at the bottom in the next forward zone. Some evolving designs included other flow patterns, such as bottom to top or side to side. These fans have been called *recirculating*, but most systems simply move the air forward in the tunnel and do not effectively recirculate the air within a zone. Only a few systems have actual or effective recirculation in any one zone of the dryer.

Existing control of a single tunnel with a constant hot air supply generally consists of controlling temperature and then modulating the air volume. Such levels of control were inadequate for preventing cracks in high-shrinkage, low-permeability products, particularly with fluctuations in weather or changes in production rates.

Changing the dryer control system to provide a constant wet bulb depression (or constant relative humidity) at the charging end of the dryer will compensate for changes in production rate as well as changes in the weather. A constant wet bulb depression or relative humidity can be maintained by adjusting the dry bulb temperature or the volume of hot air supplied to the dryer.

Controlling the wet bulb depression or the relative humidity at the charging end of the dryer gives crack-free products at the conclusion of the shrinkage stage. However, it may not satisfy the requirements for the final stage of drying, meaning that it may not provide enough temperature or time to eliminate the final amounts of pore water. This results in *shelling* and blow-up of the product in the charging end of the kiln (see Chapter 7). This lack of coordination between the initial and final stages of drying is overcome by removing much of the shrinkage water ahead of the dryer in a so-called holding or prewarming room.

The conclusion is that the old single closed tunnel concept is inappropriate for modern production with demands for production flexibility. Instead, a three-stage dryer is needed to provide for periods of prewarming, shrinkage water removal and pore water elimination. This

Modern countercurrent convection dryers have three distinct zones or stages to accomplish prewarming, removal of shrinkage water, and pore water elimination.

can be accomplished in a succession of individual dryers, but it can also be accomplished in a single tunnel by use of zone control and multiple air inlets and outlets along the length of the tunnel.

6.3 Pore Water Zone: The Final Stage of Drying toward the Exit End of the Dryer

On most dryers, the hot air supply input, called the *hot supply*, is at the final stage of drying, that is, at the point of discharge of the dry product. No shrinkage occurs in this final stage, but the product still must lose approximately half its forming water. Evaporation can be rapid during this stage since there can be no shrinkage gradient cracks. The air temperature must be high to accommodate the high thermal energy requirement for evaporating water inside the product and transferring vapor from the inside to the brick surface. The temperature required increases as the water content drops in order to overcome the energy of attraction of clay to water (as the thickness of the water film decreases). This requires temperatures of 300–500°F (150–260°C).

The hot supply upper limit air temperature is set when the product blows up in the dryer due to steam spallation. High finishing temperatures can also cause problems or thermal inefficiencies during the transfer from dryer to kiln. Room-temperature air blowing on a 400–500°F product can cause craze cracks with some raw materials. The hotter the product at the end of the dryer, the greater the heat lost during transfer to the kiln.

The lower limit air temperature in drying is indicated when the product blows up or *shells* in the charging end of the kiln. High-kaolin bodies are particularly susceptible to this problem.

Temperature elevation is required in the pore water zone, but the required dwell time at temperature is dependent not only on the temperature but also on the velocity of air, the volume of air, the tightness of setting, and the mass of product on the car. These influence not only the time for drying but also the uniformity of heat transfer throughout the setting.

First consideration is given to the source of heat. Operation and control are simpler when an independent furnace or burner provides the heat. This allows operation of the kiln with complete independence from the dryer, providing for maximum flexibility for both systems. The second choice is a constant volume of heated air (as waste heat) from the

TEMPERATURE REQUIREMENTS IN THE FINAL STAGE OF DRYING, THE PORE WATER ZONE, ARE ESTABLISHED BY:

- Time at temperature.
- Velocity and volume of air.
- Tightness of setting (ability of air to penetrate the product setting).
- Mass of product on the car.

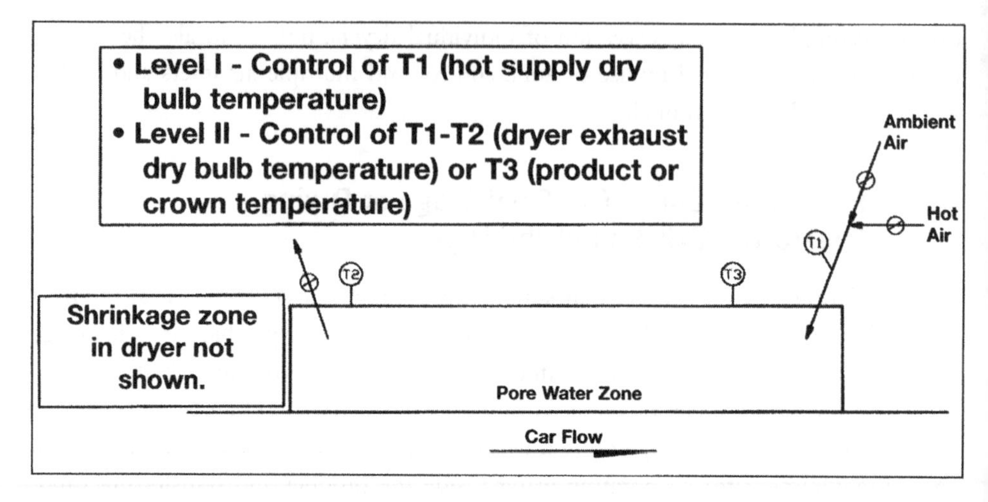

Figure 6.1: Level I and Level II controls in a countercurrent convection dryer (pore water zone only).

kiln. This causes the least upset in kiln operation and allows for reasonable kiln control. The third, and least desirable approach, is to pull variable volumes of hot air from the kiln. This can be accommodated, but it requires increased complexity and expense for kiln control. In any case, it may be necessary to supply auxiliary heaters or a means to dump excess heat to provide for fluctuating demands by the dryer.

The simplest and least expensive control can be used when there is a constant rate of charging of product to the dryer. This means consistent supply of a single size of products, consistent use of one raw material composition, and a consistent push rate (product charging rate). In this circumstance, control can be accomplished by maintaining a constant input dry bulb temperature and volume of air at the air entrance to the dryer (called *Level I control*).

Different controls applied to the pore water zone are shown in Fig. 6.1. Level I control is the least expensive and will work in nondemanding situations. The control point in Level I control is a dry bulb temperature (T1 in Fig. 6.1) in the inlet supply; that is, control is set to maintain a constant and preselected dry bulb temperature. Deviations from the set point change the ratio of ambient air to hot supply air in order to maintain the set point. Provision is made in the air discharge to utilize part of the air as supply for the shrinkage zone while dumping the excess air for other uses.

A refinement is provided by the *Level II control* system (Fig. 6.1). Here the temperature sensor (T3) is an infrared radiation pyrometer sighted on the product at a position near the zone discharge. The infrared radiation pyrometer meas-

CONTROL STRATEGY WITH LEVEL I CONTROL:

- Constant dry bulb tempefature at T1.

ures the product temperature, and the assumption is made that if the product reaches a certain temperature, it will have expelled all its pore water. Any deviation from the selected product temperature causes a change in the set point on the supply temperature. This in turn causes an adjustment in the ratio of ambient to hot supply air in order to maintain the supply temperature set point. As an alternate, a crown (roof position) thermocouple can be used, but the reading of a crown thermocouple may indicate a temperature higher than that of the actual product temperature.

There are a number of possible alternate sensors for Level II control. The decrease in dry bulb temperature from inlet to outlet of this zone (T1–T2) is an indication of the heat utilized per unit mass of air. A fundamental requirement to ensure dry product is to provide sufficient heat for evaporation of the pore water. Changing supply volume to maintain a fixed difference in dry bulb temperatures would ensure adequate heat even under conditions of varying production rates.

Maintaining a preselected dry bulb temperature (T2) at the product entrance to the pore water zone would give some measure of uniform drying conditions at the start of that zone. The dry bulb temperature could be maintained by changing either volume or temperature of the hot supply air to the zone.

The difference in the inlet to outlet dew point temperatures, T_d, indicates the water pick-up capacity per unit mass of air. Maintaining a fixed difference would ensure equivalent dewatering of the product even under varying production rates. The difference could be maintained by changing the volume of input supply air or its temperature.

CONTROL STRATEGIES WITH LEVEL II CONTROL:

- Product exit temperature (T3).
- Decrease in dry bulb temperature (T1–T2).
- Constant dry bulb temperature at the product entrance (T2).
- Difference in dew point temperatures at locations for T1 and T2.

Level III control (Fig. 6.2) adds one or two recirculation zones in the pore water zone. This permits greater control in shaping of the drying curve between the inlet and outlet dryer positions. Recirculation provides for greater uniformity of drying throughout the setting and also greater output through a given length of tunnel. The dry bulb temperature (T1) at the supply inlet and in each recirculating zone provides the signal for control. These temperatures are maintained at preselected set points by varying the ratio of hot supply air to recirculated air at each location. A dryer with two recirculating zones might use preselected temperatures of 280°F (138°C) for T4, 340°F (171°C) for T5, and 400°F (204°C) for T1. A key to this system is the provision to introduce hot supply air at three different locations.

Level IV control offers the same options as Levels II and III except this time an auxiliary burner is added to supplement heat from the kiln or as the primary source of heat. Again, modulating volume or temperature provides the needed control action.

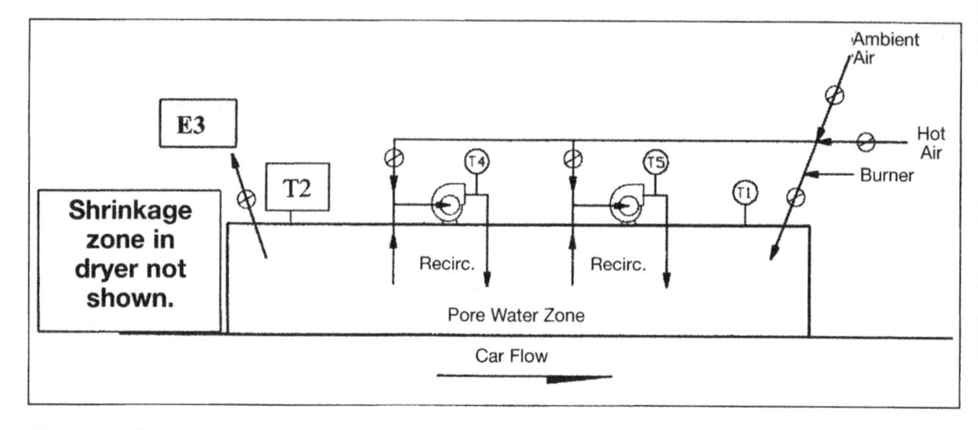

Figure 6.2: Level III and Level IV controls in a countercurrent convection dryer featuring two recirculating zones (pore water zone only).

Exhaust from the pore water zone is optionally controlled by a pressured damper (E3). A fixed level of pressure is maintained. The exhaust from the pore water zone can be used as the supply source for the shrinkage zone and the prewarming zone. Any excess over the demands of these zones would need to be dumped or applied to other uses.

6.4 Controlling the Shrinkage Water Stage (Product Entrance End Zone of the Dryer)

The maximum rate of evaporation is limited in the shrinkage stage to the rate at which water can flow from the interior to the surface of the object. Exceeding this maximum rate will produce a water gradient accompanied by a shrinkage gradient that, when excessive, will cause the product to crack. Materials differ widely in their potential maximum rate of internal flow. Montmorillonitic clays and other highly plastic materials with a high content of colloidal particle sizes have slow rates. Non-plastic shales and highly grogged bodies will have high rates of water movement. Kaolin bodies will have high rates if their state of agglomeration is high.

The rate of evaporation can be controlled by the difference between dry and wet bulb temperatures (i.e., controlling the relative humidity), or the rate can be controlled by the difference in dry bulb and dew point temperatures. The wet and dry bulb differential is the most common method of control. Relative humidities for air with difficult-to-dry materials may be 85–95% in the shrinkage zone.

Reasonable efficiency requirements of the dryer dictate that the exhaust should be between 108 and 120°F (42–49°C) dry bulb temperature with a relative humidity of 85–95%. Lower humidity indicates paying for air movement that is not doing its job of water hauling. Lower temperature indicates paying for excessive quantities of air.

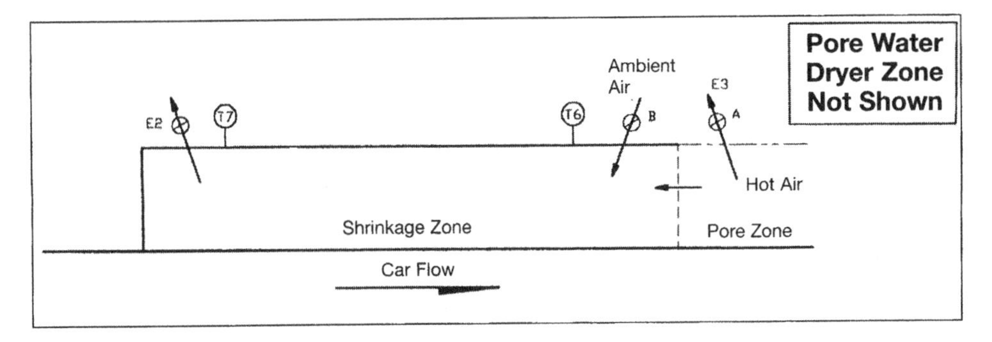

Figure 6.3: Controls in the shrinkage water zone for Level I and Level II control.

The charging end drying environment is selected to suit the product properties and also to provide the desired economy of operation. The length of the shrinkage zone must be sufficient to complete shrinkage for the particular material, setting pattern, and charging rate. The rate of evaporation can be gradually increased with increasing length into the shrinkage zone. Evaporation rates are abruptly increased on entry of the product into the pore water zone.

6.5 Basic Shrinkage Zone Control

Air serves two distinct functions in convection dryers. Heated supply air provides the necessary thermal energy for heating up the ware and evaporating water. Also, air supplies the conveying mechanism for sweeping water vapor out of the dryer. The two functions are not equivalent and the quantity of air for heating will be different than the quantity of air for conveying, except for one operating condition where the two come into balance. This balance exists only for a single supply temperature for one exhaust temperature and humidity and for one rate of loading. This is the limitation of the single tunnel concept designs of the past: only one inlet of hot supply air and one exhaust of wet air. Flexibility of operation and control requires multiple air inlets and outlets and the ability to isolate each drying stage.

Level I controls, applied to the shrinkage zone of a dryer, are shown in Fig. 6.3. The discharge from the pore zone is the hot supply air source for the shrinkage zone, but in most cases the pore zone discharge air properties exceeds the energy requirements of the shrinkage zone. The extra heated air is exhausted (E3) and the remaining heated air continues down the tunnel to the shrinkage zone.

Level I control sets the dry bulb temperature (T6) to a preselected temperature. This level is maintained by the percentage opening of Dampers A and B. Opening of Damper B bleeds in cool ambient air to reduce temperature. The simultaneous opening of Damper A reduces the quantity of hot air from the pore zone.

Thus, changing the ratio of ambient to hot supply air can maintain the dry bulb temperature in the shrinkage water zone air input at a desirable set point.

Level I control will work for one rate of input loading, but will not accommodate variations in input rate. Level II control will accommodate changing input rates. This control is based on maintaining a fixed evaporation rate at the charging end of the zone. This constant rate of evaporation is accomplished by maintaining a fixed difference between the dry and wet bulb temperatures (relative humidity) at the charging end of the zone (T7). The difference between dry and wet bulb temperatures is called the *wet bulb depression.*

Deviation from the preselected set point in Level II control changes the set point on the dry bulb temperature at (T6). This in turn calls for changes in the opening of Dampers A and B to provide the required temperature. This level of control maintains a fixed air volume through the shrinkage water zone.

An alternate method of Level II control is a variable volume method. This can be accomplished by either a damper at the exhaust (E2) or by use of a variable-speed exhaust fan. The quantity of air can be reduced by either closing the damper or by reducing fan speed. The flow rate can be adjusted to give the preselected relative humidity. The dry bulb temperature (T6) can be independently controlled, as stated earlier.

A further refinement in control is the Level III control shown in Fig. 6.4. This level provides for recirculation of part of the dryer exhaust to the discharge section of the shrinkage zone, increasing control of the drying rate throughout the length of the shrinkage zone. The relative humidity at T7 controls the ratio of exhausted air to recirculated air. The dry bulb temperature (T6) continues to be controlled by blending ambient air with heated air by adjusting Dampers A and B.

6.6 Fan Transfer between Shrinkage and Pore Water Zones

The previous examples of levels of control provided one hot supply fan and one exhaust fan for the one zone in the dryer. Control sensitivity can be improved by adding a fan to transfer heated air from the pore zone to the shrinkage zone (Fig. 6.5) to achieve Level IV control. Part of the heated air will still drift along the tunnel, but the portion of air transferred by the fan will break up crown drift as well as improve control quality.

Level IV control, like Level II control, is based on maintaining a fixed relative humidity at the charging end (T7). This, in turn, controls the set point on dry bulb temperature (T6). The temperatures are maintained by controlling the opening of Damper A with simultaneous adjustment of the opening of Damper B. The adjustments to Dampers A and B change the supply temperature without changing the volume flow rate to the shrinkage zone. The relative humidity at the charging end (T7)

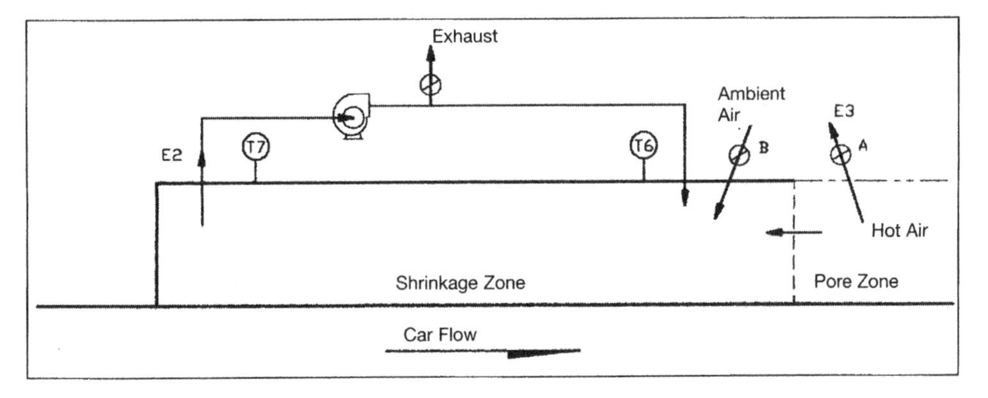

Figure 6.4: Controls in the shrinkage water zone for Level III control.

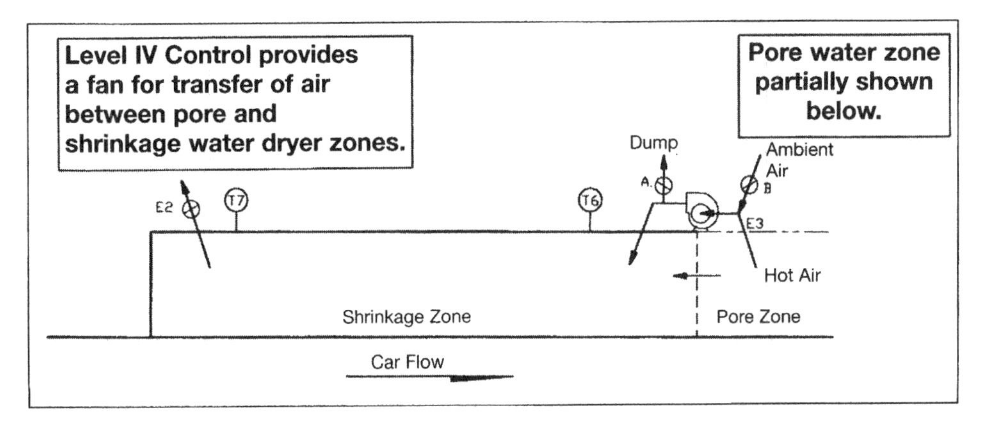

Figure 6.5: Level IV control shown in the shrinkage water zone.

could likewise be controlled with volume airflow. Changes in the opening of Damper A would increase or decrease the volume flow rate in the shrinkage zone, thereby increasing or decreasing the relative humidity at T7.

6.7 Time Lag

Time lag is a problem in all dryer control systems. There may be a delay in the control action because of the time required for the sensing element to fully detect the change in the variable being sensed. For example, it takes time for a thermocouple to respond to a temperature change when it is surrounded by primary and secondary protection tubes.

There may be time delay in the process itself, as illustrated in Fig. 6.3. Any

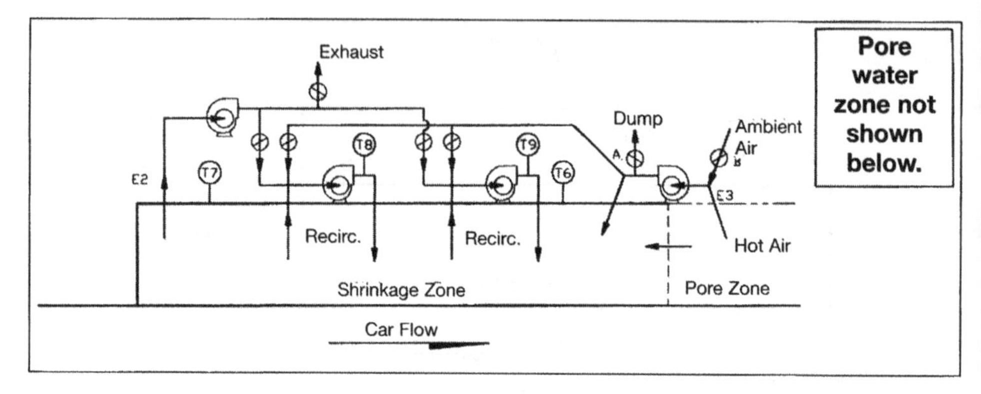

Figure 6.6: Recirculation in a dryer shrinkage water zone.

deviation in exhaust temperature results in an immediate adjustment to Dampers A and B; however, the result of this adjustment will not become apparent at T7 until sufficient time has passed for equilibrium along the entire length of the shrinkage zone. The change at T7 will then occur gradually as products over the entire length of the zone come to equilibrium with the new environment. This means the damper adjustments tend to overcorrect, therefore some system is required for limiting the control correction.

Time delays can be programmed into some control systems. Time delay problems can be reduced by the addition of recirculation zones within the dryer section.

6.8 Recirculation Systems

A single tunnel can be controlled in separate zones by use of recirculation, which takes air from a given area in the dryer length and inputs the air back into the dryer at a point one or two car lengths further into the dryer, against the air flow. Additions of hot or moist air to a recirculation system allow control of temperature and humidity.

Recirculation not only provides a means of individual zone control, but also increases the rate of heat transfer. The increased velocity resulting from recirculation increases the air's ability to penetrate into denser product settings and provides for greater uniformity throughout the set. Recirculation, combined with separate inlet and outlets for each zone, permits the drying schedule to be shaped along the length of the tunnel, providing for maximum output. Abrupt changes can be made from zone to zone, rather than the gradual transition available in a tunnel dryer without recirculation.

Two stages of recirculation are added in the shrinkage water dryer zone shown in Fig. 6.6. The wet bulb and dry bulb temperatures are controlled in each recirculating stage, thus providing a more immediate response to changing conditions.

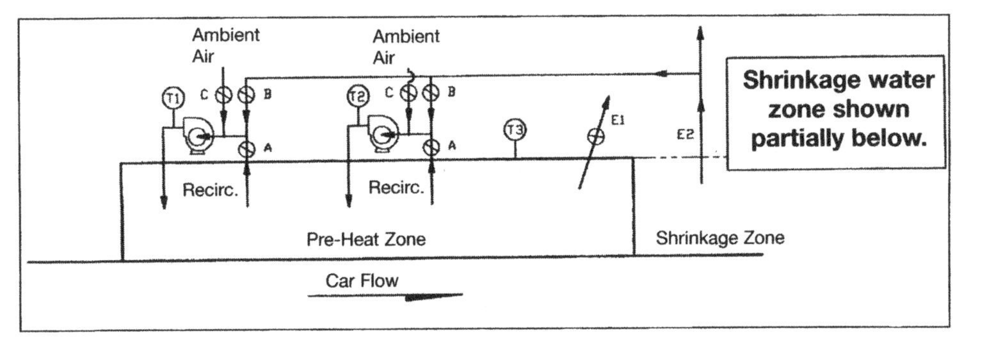

Figure 6.7: Preheat zone of a countercurrent convection dryer.

Control is achieved in each stage by proportioning the amount of hot supply air (C) and wet dryer exhaust air (D) to each recirculation zone. The hot supply source is the exhaust from the pore water section. The dry bulb temperature at T6 is controlled by adjusting Damper B and the relative humidity (T7) is controlled by Damper A.

The use of the recirculating zones permits shaping of the drying curve along the entire length of the shrinkage zone. This provides for greater efficiency and greater output within a given length of tunnel.

6.9 Preheat Stage

Economy of operation requires the major dryer exhaust to be at a temperature 100–120°F (38–49°C) and at 90+% relative humidity. This means that the dew point temperature of the exhaust will be 90–110°F (32–43°C). The product temperature may be below the exhaust dew point temperature, particularly during the winter. A ~5°F (~3°C) deficit is probably inconsequential, but larger differences can cause condensation on the product with consequential slumping, cracking, or scum development. It is necessary to preheat the brick above the dew point temperature to avoid these defects. The preheating must be accomplished without exceeding the permissible drying rate of the material. Thus, humidity must be controlled along with heating.

There are systems for accomplishing preheating. The dryer exhaust cannot be used alone for preheating or in direct contact with the incoming product, because exhausting from the charging end of the preheat zone at reduced temperatures would require condensation of moisture from the air. Of course, the dryer exhaust could be adjusted downward to 50°F (10°C) and 90% humidity in the wintertime, but this would require large quantities of air and reduced dryer output and the quality of the control would be poor.

Controls on a preheat tunnel attached to the shrinkage zone are shown in Fig. 6.7. Two or more recirculating zones can be used in the preheating zone. Note the

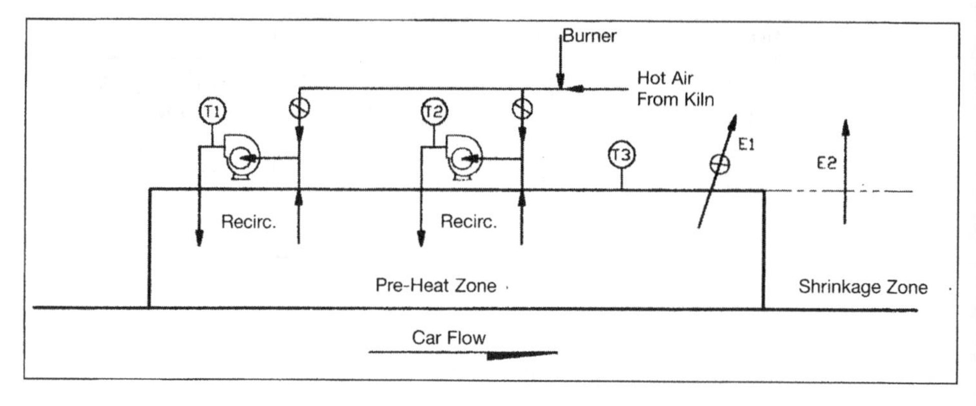

Figure 6.8: Preheat zone using kiln exhaust as the hot supply air.

reversal of airflow in the preheat zone. The air flows with the direction of the car travel, rather than countercurrent as in the remainder of the dryer.

To control the preheat zone, exhaust (E2) from the shrinkage zone is used to supply heat as well as humidity. It is necessary to blend the dryer exhaust with ambient air in order to drop the dew point temperature to an acceptable level. The desired dry bulb temperatures (T1) and (T2) are achieved by controlling the ratio of dryer exhaust air (B) to ambient air (C). The wet bulb depression or relative humidity (T3) is controlled by an increase or a decrease of the air pulled from the preheat tunnel (E1).

Returning the preheat section air to the main dryer exhaust fan (E2) would result in variable volume through the shrinkage zone. To avoid this, it is necessary to have a separate exhaust fan for the preheat section or a variable speed main exhaust fan (E2) that could be adjusted to accommodate the amount of exhaust diverted to the preheater section.

The use of shrinkage zone exhaust air in the preheating section may result in condensation on blending of ambient air with the exhaust air. This possibility could be avoided by replacing the dryer exhaust air as a heat source with hot supply air from the kiln or a separate burner (see Fig. 6.8). In the case of using kiln exhaust as the hot supply, the desired dry bulb temperatures of T1 and T2 are achieved by controlling the amount of hot supply air added to each recirculation fan. The wet bulb depression or relative humidity (T3) is controlled by increasing or decreasing the air pulled from the preheat tunnel (E1).

For installations where large changes in loading are commonplace, a control system using both hot supply air and dryer exhaust air is shown in Fig. 6.9. Two recirculating zones are shown, but others could be added. The wet bulb temperature or relative humidity (T3) could be controlled by adding more or less dryer exhaust air (B) or changing the preheat exhaust volume (E1). The dry bulb temperatures T1 and T2 could be controlled with adjustments to damper A.

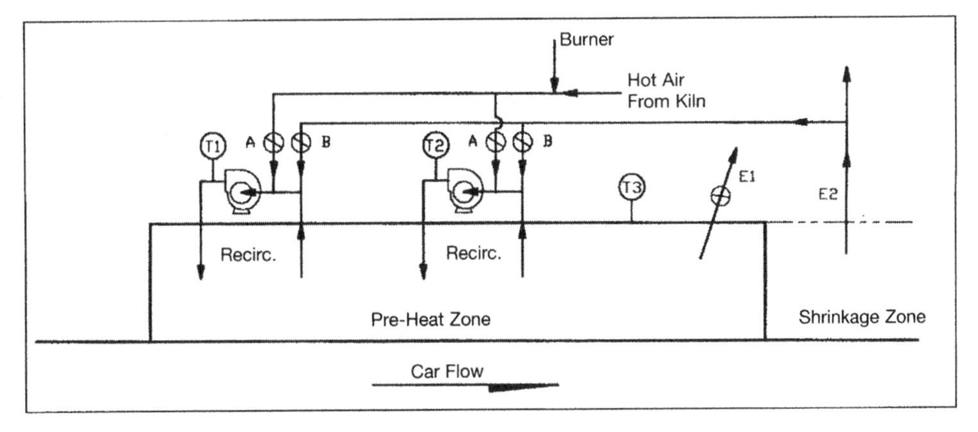

Figure 6.9: Preheat zone using both kiln and dryer exhausts as the hot supply air.

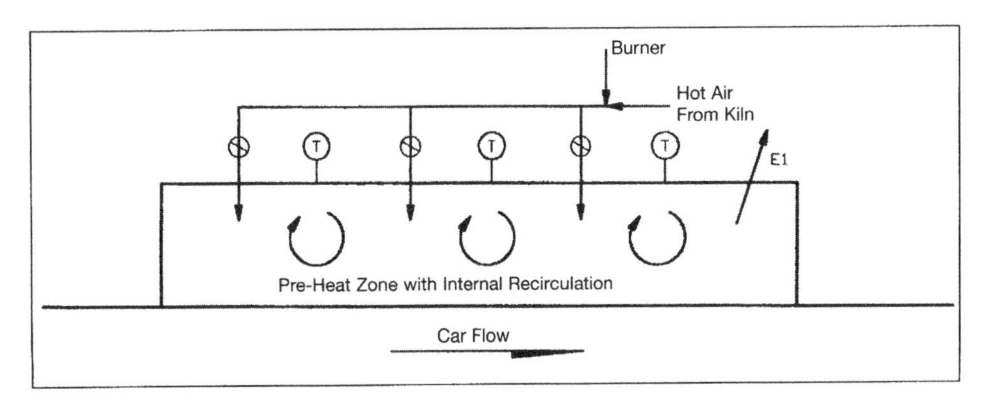

Figure 6.10: Use of a holding room for preheating.

Another system of preheating (Fig. 6.10) is to place the green product in a closed room or tunnel (often referred to as a holding room) with restricted air circulation. The drying product builds up the humidity in the confined space. The product temperature is gradually increased by increasing heat, such as by using hot air from the kiln or by using a direct-fired burner. Relative humidity is altered by changing the preheat exhaust air volume of E1. For raw materials that are more difficult to dry, internal air recirculation should be provided in the preheating room by using oscillating fans, pulsing fans, or low-velocity ceiling fans. All preheating systems should be operated with positive pressure to eliminate cold-air leakage and localized condensation. Preheat exhaust fans are not always necessary.

Some brick bodies, for example, require exposure to a gentle drying environment immediately after setting. These brick may crack when exposed to the slightest

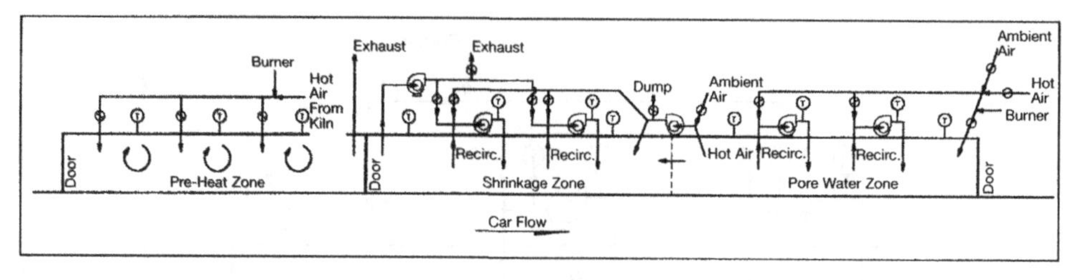

Figure 6.11: Composite tunnel dryer showing all zones in line.

draft in the building and should be protected at all times. Brick are weakest at their maximum water content and show marked increase in strength as soon as 1% water is removed. Furthermore, if the product does not start drying, its strength decreases with time. Brick under load gradually creep to failure (cracking). Some brick bodies may be able to resist cracking for up to 6 h, but will gradually deform to failure in 12–24 h. Brick bodies of this type should be moved immediately to a preheater controlled with a high humidity (90%) and relatively low temperature. This is particularly important for the brick under load at the base of the setting.

6.10 The Composite Dryer

A complete dryer using many of the control options discussed in this section is shown in Fig. 6.11. Although each zone appears to be the same length, they will differ depending on production schedules and raw materials. Preheating systems generally enclose all materials between the setting area and the main dryer entrance. Lengths of the shrinkage zone and the pore water zone must be based on the individual product requirements, including raw material composition, particle size, and product size. The challenge remains in designing a dryer and controls to match the requirements of the raw material and the flexibility of production schedules.

6.11 Problems

1. You are designing a pore water (Stage II) section of a countercurrent tunnel dryer analogous to the one shown in Fig. 6.1 given the following set requirements:

   Tunnel cross section (freeboard above car deck): 10 ft height × 10 ft width (rectangular dryer tunnel with a flat suspended crown).

   Tunnel length: 50 ft.

   Hot supply conditions (waste heat): $T = 450°F$ and $T_w = 106°F$.

   Product density along tunnel allows 35% void area including freeboard.

   Minimum air velocity desired in void and freeboard area = 15 ft/s.

The minimum air velocity requirement was determined from experience as that required to achieve good heat transfer in this type of dryer tunnel.

(a) What is the minimum airflow, in pounds of air per minute (DB) at the hot supply, to meet the minimum linear velocity required at the entrance end of the dryer? Assume uniform air flow from the top to the bottom of the tunnel.

(b) Assume the hot supply air loses 45% of its enthalpy in passage along the tunnel from the hot supply to the pore water zone exhaust (at T2). This results in a heat loss by the air to the product (assuming zero dryer loss). What is the minimum air velocity at the hot supply to create the minimum linear velocity in the dryer tunnel at the zone exhaust?

(c) What is the zone exit air dry bulb temperature?

2. Given ambient air at T = 65°F and relative humidity (H_r) of 50%, what volume of ambient air in pounds per minute should be mixed with waste heat to form the hot supply for the dryer of Problem 1 where the new hot supply entrance requirements are T = 325°F and the mass flow must be 1500 lb/min at the exit of the pore water zone?

3. Given the pore water zone (Stage I) of a dryer designed as in Fig. 6.5. The input of this dryer is as follows:

Shrinkage water zone output:

T = 270°F
H_r = 0.03%
volume = 30,000 cfm

Ambient air:

T = 80°F
T_w = 60°F

Shrinkage zone output:

T = 120°F
H_r = 85%

If the material being dried can tolerate a zone input temperature of 175°F, answer the following:

(a) What is the maximum product input rate to the dryer if the average product weighs 4.5 lb (green) at a moisture content of 16% (WB)?

(b) At the maximum product input rate, what are the mass flow rates of waste heat and ambient air making up the zone input?

4. Given the dryer of Problem 3, with winter conditions as follows:

Ambient air:

$T = 35°F$
$H_r = 5\%$

(a) What is the maximum product input rate given the same product data?

(b) If a hot air burner is added to the system to increase winter production levels to the same as those possible in the summer, how many Btu/lb of dry air must be added at the zone input in the winter?

5. What is the thermal efficiency of the dryer in Problem 3?

6. What is the air efficiency of the dryer in Problem 3?

7. Why is the airflow in a holding room usually concurrent with the direction of product travel while it is countercurrent to the direction of the product movement in the dryer?

8. Given a holding room as described in Fig. 6.10 with the additional data shown below:

Hot air from kiln:

$T = 350°F$
$H_r = 0.05\%$
Available airflow = 2000 cfm

Ambient conditions:

$T = 60°F$
$H_r = 20\%$

Holding room dimensions above car deck: 13 ft high × 10 ft wide.

Product density along tunnel allows 35% void area including car cross section and freeboard.

Minimum air velocity desired in void and freeboard area = 1 ft/s.

What airflow is required of kiln exhaust and ambient air for the holding room entrance condition to achieve a holding room exit condition of $T = 80°F$ and $H_r = 95\%$?

9. Given the product characteristics of Problem 3, how many bricks per hour can be sent through this holding room?

10. **(a)** Discuss how the holding room should be controlled, assuming five days of production (forming) at one shift of forming per day and assuming that enough cars must be maintained in the preheat room over the weekend so that the minimum exit conditions are maintained in the dryer given in Problem 8.

(b) Discuss the criteria used in selection of fans used in the recirculating section of a dryer in a brick plant.

11. How do the criteria differ for fan selection for holding rooms in a brick plant?

12. Discuss the potential problems with using undercar air from the kiln as opposed to using waste heat from the cooling section of the kiln with respect to scumming starting in a holding room.

13. Discuss potential maintenance to a dryer superstructure in a coal fired brick plant.

14. Discuss measurement of airflow in a dryer using a Pitot tube (see www.epa.gov).

15. Discuss the potential use of handheld wind speed measurement devices in balancing airflow in a holding room.

16. In a multiple track holding room, what techniques could be employed to direct air more efficiently through the burden on a dryer car instead of around the car?

17. Devise a test strategy to determine if cracks in fired bricks on the bottom of a product setting originate in the kiln (as preheat cracks) or in the dryer.

18. Discuss strategies to increase green strength of brick products to avoid undue deformation or cracking in the bottom courses of the dryer setting.

CHAPTER 7

Drying Defects and Drying Shrinkage

7.1 Introduction

Many cracks that are blamed on the dryer are actually the result of raw material selection, mixing procedure, forming procedure, or product design. Drying schedules are powerless to correct these faults, whereas correcting the source of the problem can lead to vastly accelerated drying schedules.

Cracks and defects that are affected by the drying schedule have been known since the early days of processing of ceramics. These include *dryer cracks* — a term generally applied to cracks resulting from excessive drying rates and shrinkage in Stage I of drying. These dryer cracks may also develop because of excessive loading of the product in the green state; that is, the product cannot support the weight of the hack. The products in the bottom of the setting may also experience cracking because of their shrinkage, relative to the lack of shrinkage of the underlying refractory car deck block.

In the extreme, excessive heating rates can lead to *steam spallation* in the product, where the product explodes due to internal steam pressure. *Shelling* is spallation of an edge or face when residual moisture in the product is inserted into a kiln or hot dryer.

Scumming is the result of acid gases originating in the kiln coming into the dryer with the waste heat and causing discoloration in the fired product. In this case, the problem occurs in Stage I of drying or in the predryer when the wet surface of the product becomes slightly acidic, thus rendering salts near the surface soluble and mobile. The resulting deposit typically causes a color change in the fired product.

PART A: DRYING CRACKS RESULTING FROM MATERIAL PREPARATION

7.2 Shrinkage Gradient Cracks

The most common source of cracking is the ceramic part's surface attempting to shrink before the interior begins to shrink. This produces cracks called *craze*

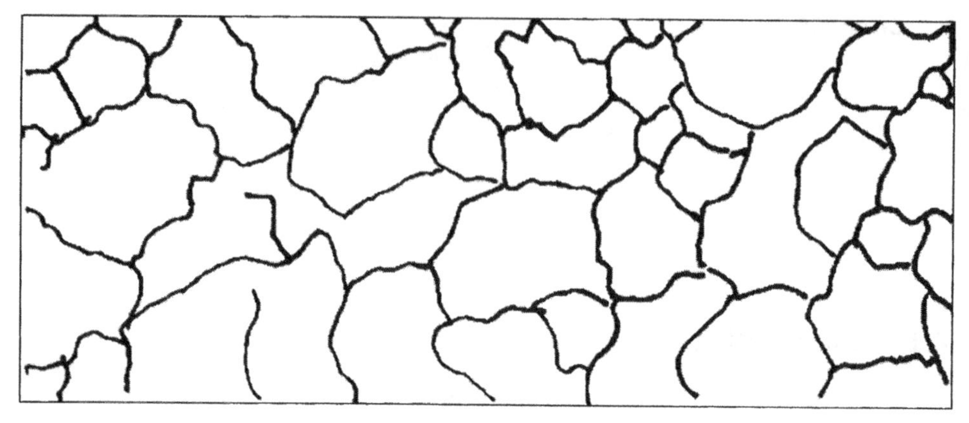

Figure 7.1: Surface craze cracking on a brick specimen of 15% shale mixed with wet clay. (Tracing from a copy. Original photograph by Gil Robinson not available.)

cracks or *mud cracks* (Fig. 7.1). Fast drying, low permeability to internal moisture movement, and high shrinkage, singly or in combination, favor this type of cracking. The classic cure is slower drying or making the object more resistant to cracking by adding grog, shale, or other nonplastics. These nonplastics cut drying shrinkage and make the body more permeable. However, it is possible that this cure creates new and/or other problems in the product.

This latter situation was found by attempts to speed up the drying of a "tight" plastic clay used by a Mississippi brick manufacturer. Shale was admixed in amounts of 15, 30, and 45% into the body. It was found that none of the mixes would tolerate accelerated drying, including the mixture containing 45% of nonplastic shale. In fact, it was found that this mixture could not stand even very slow drying in static room temperature air. This was in contrast with the 100% clay composition that did dry without cracking in a static room environment.

An effort was made to determine the reason for the failure of this traditional approach to a drying problem. Examination of the specimen dried in static air showed the appearance of shrinking islands dispersed in a sea of coarse aggregate particles. It was found that the specimen formation procedure mixed wet plastic clay with damp crushed shale. The clay did not blend uniformly with the shale, but rather remained agglomerated. The mixing procedure was at fault, rather than drying environment.

New specimens were prepared by drying the clay and powdering it prior to mixing with the dry sized shale. Water was added to the mixture and specimens were formed. This procedure stopped the cracking of the 45% shale composition even under accelerated forced-fan convection. Now, even the 15% shale composition resisted accelerating drying schedules with no problem. Correcting the mixing procedure resulted in much faster drying schedules for the 15% shale composition than could be achieved by the clay alone.

7.3 Nonhomogeneity in Structure

Agglomerates or granules may be purposely prepared as the feed for the forming process. The persistence of their structure can be a source of cracking. This was illustrated with trials made from the factory-extruded brick of the tight plastic clay discussed above. It was decided to reconstitute the material into half-scale solid brick to increase the number of trials that could be made from a full-size brick. Furthermore, it was desirable to determine if laboratory-prepared brick would imitate the drying faults obtained by commercially produced units. The plastic brick were cut up with a knife and the pieces were placed in a die and pressed into a half-scale solid brick.

The plant-extruded brick cracked when placed in front of a 24 in. fan in the room environment. The laboratory-prepared specimens showed gross cracking under the same conditions. Also, the laboratory specimens showed cracking in the room environment without the fan circulation. Factory-produced brick did not crack under these conditions. The pieces of clay were mashed together but did not knit or homogenize sufficiently to provide strong interparticle bonds.

Cracking occurs in the weakest portion of the body, and the intergranule zone provided the weak link in this case. Attempts were subsequently made to heal together these shreds by conditioning under high humidity. Plastic specimens were aged overnight while wrapped in aluminum foil. These preconditioned specimens were then subjected to forced-air and static room–air drying. The preconditioning reduced the severity of cracking compared with nonaged samples. In Fig. 7.2, forced-air drying of brick at ambient laboratory temperatures caused only minor cracking on an aged specimen constituted from shredded material. By contrast, the aged specimen on the right exhibits more extensive cracking.

Aging in the plastic condition increases the adhesion between particles and reduces the severity of cracking; however, it does not completely eliminate the structural weakness or cracking.

It was thought that interparticle adhesion could be improved by dampening the plastic shreds prior to pressing. Instead, this procedure increased cracking compared to nonwetted shreds (Fig. 7.2). It was anticipated that the use of higher pressure during forming would eliminate the weakness zones between pieces; however, the shreds obtained from the factory-extruded brick were so stiff that a reduction in cracking was not obtained. A specimen was pressed with such force that the material extruded out the sides of the die resulted in cracking even when it was slowly dried.

Changing the preparation procedure changed the cracking results. Drying the factory-extruded brick and crushing it to −14 mesh prior to mixing with water and

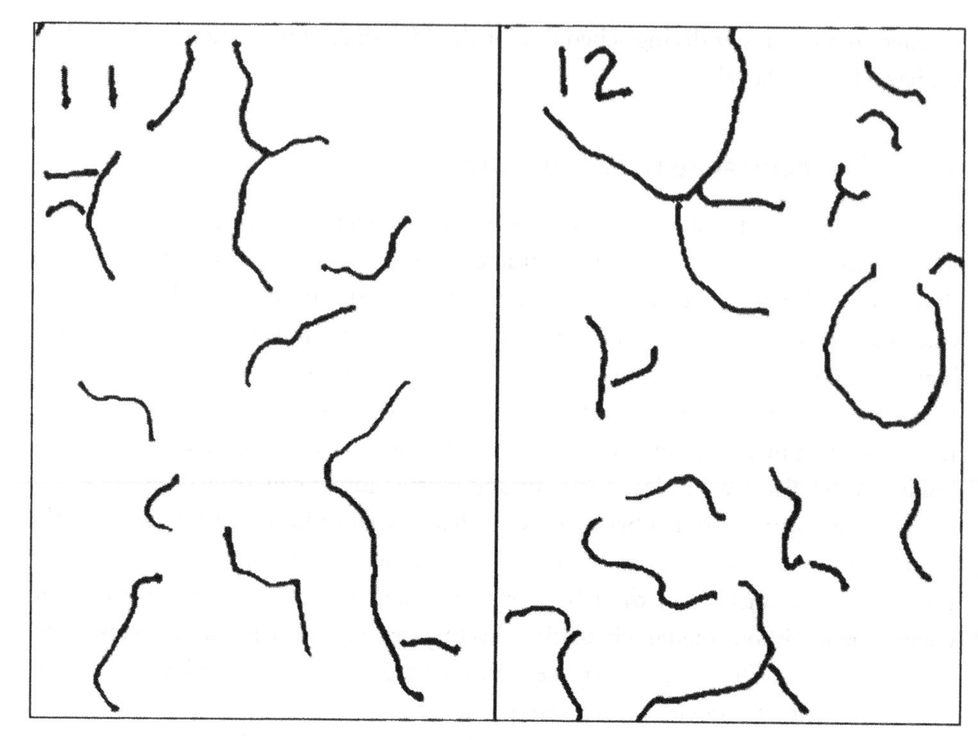

Figure 7.2: Effect of forced air drying on aged specimens: Surface cracks formed at agglomerate edges. (Tracing from a copy. Original photograph by Gil Robinson not available.)

forming produced crack-free specimens. The units were crack-free under both static room temperature drying and with the accelerated fan drying. The softer granules resulting from this preparation procedure produced homogeneous samples that were well knit together. The preparation procedure corrected the fault in structure and permitted accelerated drying schedules.

It appears that any agglomerate type will persist as an identifiable structural element in the formed product when the agglomerate is stronger than the applied forming force. This results in an interparticle weakness that may produce drying cracks. Agglomerates that are destroyed by the forming force do not produce drying cracks. Agglomerates may come from recycling scrap plastic brick, addition of crushed dry brick, and clay raw material lumps.

The powder-pressed specimen did not crack under fan drying, but the plant-extruded brick cracked. The difference could have been from structural flaws in the factory-extruded brick, but it was also noticed that the laboratory brick contained slightly less water.

The influence of water quantity on cracking was evaluated by preparing powdered clay specimens with increasing forming water. A specimen with 14.5% water showed no cracks when fan dried with room air. A similar specimen with 15.7% water showed slight cracking and a specimen with 16.7% water showed pronounced cracking. This suggests that small changes in forming water can have a pronounced influence on the occurrence of drying cracks. The laboratory-produced specimens with 16.7% water exhibited the same drying behavior as the factory-extruded brick.

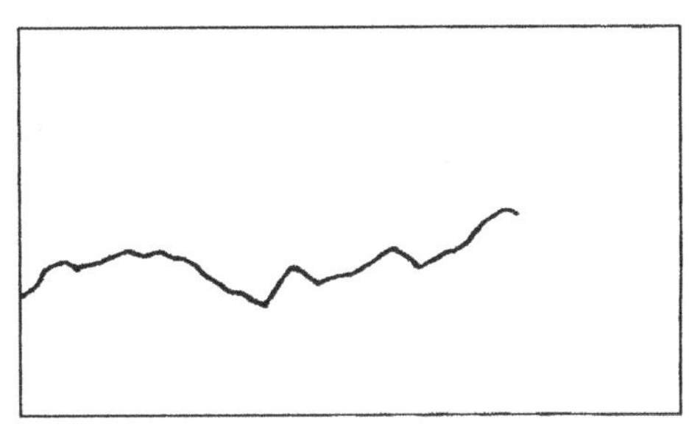

Figure 7.3: Cracking on the low moisture content side of a composite clay specimen dried in still room-temperature air. (Tracing from a copy. Original photograph by Gil Robinson not available.)

7.4 Water Gradient from Nonuniform Material

Cracks are likely when there is a change in water quantity from one position to another within an object (Fig. 7.3). Cracks are shown on the face of the brick molded with 16% water. The molding was accomplished by mixing two batches of powdered material with different water quantities. One batch was prepressed to a half-brick thickness. The upper surface was roughened with a brush without removing any material from the mold. The second batch of higher water content was added and the pressing of the unit was completed. The specimen was dried in still room-temperature air.

It is interesting that the lower water content side of the specimen exhibited the cracking. The lower water content section was sufficiently dry to be nearly elastic in its behavior. The 16% water content face was at stiff extrusion consistency or plastic in behavior. It is likely that the deformable high water content side relieved the strain difference without cracking. Continued drying caused the high water content side to shrink and produced convex bowing on the low water content side. This caused or further aggravated the cracking.

7.5 Water Gradient from Drying Gradient

Blowing air on one face of a brick will dry that face faster than one exposed to little or no airflow. This will produce a water difference on one side and a consequent shrinkage gradient within the brick. Cracking will result when the stress from shrinkage gradient exceeds the strength of the object. Cracks will appear as craze cracks on the face directed toward the incoming airflow.

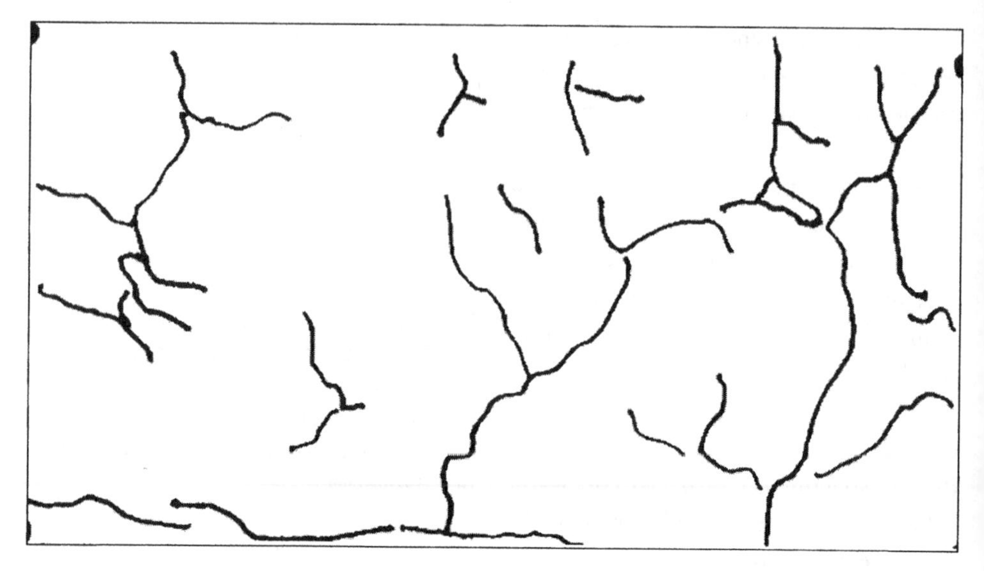

Figure 7.4: Cracking caused by air impingement in drying. (Tracing from a copy. Original photograph by Gil Robinson not available.)

The pattern of cracking will change when the object shape, coupled with drying environment, causes warping of the object. For the specimens shown in Fig. 7.4, air at 230°F (110°C) was directed on one face of a specimen. This produced cracks within seconds on the windward face followed by bending of the specimen toward the wind source. These windward face cracks closed as the bending deformation progressed, and cracks appeared on the opposite face. The concave curvature of the windward face stretched the back or leeward face in tension and produced cracks. The more severe cracking of the leeward face persisted in the dry specimens.

Slower drying rates were applied to specimens similar to those in Fig. 7.4 by blowing room-temperature air on one face of each specimen. Cracking was worse on the sides opposite to those exposed to the air flow. Increasing water content in these specimens increased the amount and severity of cracking. Cracks were observed on the side of brick specimens in experiments similar to those where face cracking was produced.

7.6 Forming Flaws and Drying Cracks

The direction of warping may change depending on the speed of drying, the specimen shape, and the extent of initial cracking. In an extruded specimen made from a relatively nonplastic raw material, convex curvature was observed on the face exposed

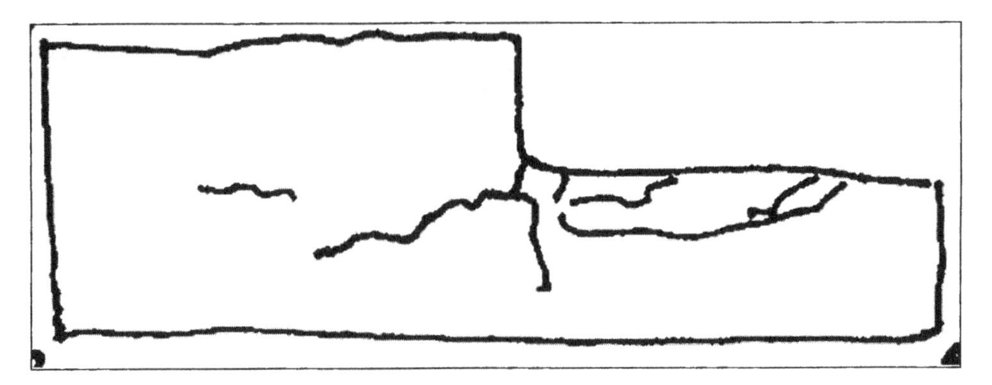

Figure 7.5: Effect of thickness changes on cracking in drying of an L-shaped piece. (Tracing from a copy. Original photograph by Gil Robinson not available.)

to an air blast at 230°F (110°C). In other words, the cracking was extensive on the windward side and almost nonexistent on the leeward side.

The cracks were found to enter from the edges of the specimen, and they occurred in weak zones between the spiraling auger discharge. It was found that the use of a longer shaper cap with a lower entrance angle permitted faster drying without cracking. This example indicates that correction of forming flaws can eliminate cracking that may have been otherwise attributed to drying speed.

In further experiments, double specimens made with two water contents were dried rapidly by directing room-temperature air toward one face. Combining water gradients from air flow and material composition showed the influence of air flow to be more significant than the influence of material.

7.7 Thickness Gradients

Differences in specimen thickness can cause cracking (Fig. 7.5). This specimen was prepared with 16% water. The single specimen showed no cracking when dried by a fan with room temperature air, however, a second sample was cut to leave one-half thickness for half of the length of the specimen. Fan drying this specimen caused cracks in the thick section and splitting cracks between the thin and thick sections.

There is more water to remove in the thick section. The same evaporation rate is used for both sections, so the shrinkage water removal is completed in much less time from the thinner section. This means that there is a shrinkage difference between the thick and the thin sections. This dimensional difference creates a cracking force between the sections.

PART B: THEORETICAL TREATMENT OF DRYING SHRINKAGE

7.8 Introduction

Drying shrinkage can be treated in a theoretical fashion by developing a series of equations linking the response of a material to a developing shrinkage gradient. There are obvious limitations to this approach because many of the material parameters are unknown, and it can be argued that a drying ceramic also exhibits a plastic response if its water content is high enough (as shown in the preceding section). The model has value, however, in helping people visualize the drying process.

If a semi-infinite solid of length $2w$ experiences drying only from its ends, a series of moisture gradients develop during drying. These can be illustrated as in Fig. 7.6.

The assumption is made that the moisture movement process is described by Fick's First Law, which is also commonly used to describe diffusion processes. In this law, the flux of molecules, J, is related to a concentration gradient for moisture, dm/dx, where m is the moisture content and x is a linear dimension.

$$J = -D\frac{dm}{dx} \tag{7.1}$$

Equation 7.1 describes a flux in one arbitrary dimension, and to the approximate the moisture movement in three dimensions, a factor of 3 can be inserted into equation. Defining m_{surf} and m_{avg} as the moisture concentration at the surface of the drying object and the average moisture content of the green piece, respectively, we have:

$$J = -3D\frac{m_{surf} - m_{avg}}{w} \tag{7.2}$$

Rearranging, we have:

$$m_{surf} - m_{avg} = -\frac{1}{3}\left(\frac{Jw}{D}\right) \tag{7.3}$$

Equation 7.3 relates the moisture gradient to the drying rate (or flux J), the half-thickness of the specimen, and the rate of moisture movement. In order to complete our analysis, it is necessary to make an analogy between the drying process and

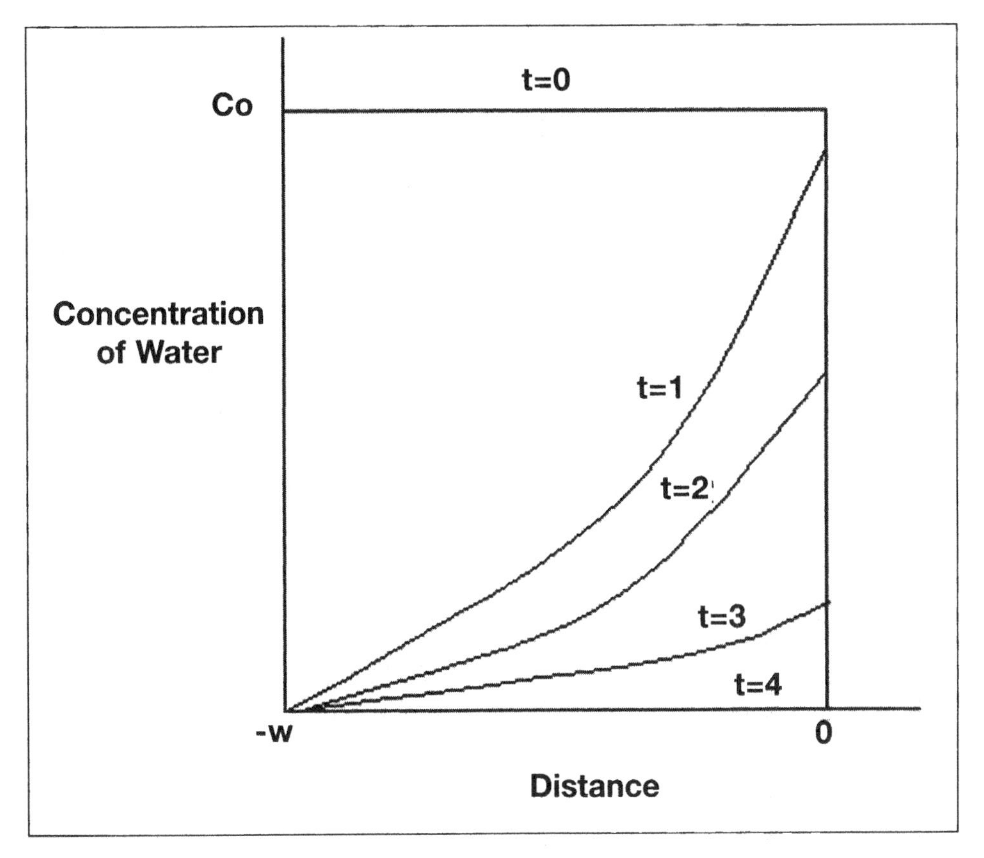

Figure 7.6: Semi-infinite solid drying model (showing half length or the left side).

a heat transfer process or thermal shock for the same type of specimen. The rationale for doing this is that both processes result from a gradient. In the case of drying, differential stress is created by a shrinkage gradient. In the case of thermal shock, a surface stress is created by a temperature gradient when a specimen is quenched (rapidly cooled).

From thermal shock theory, we know we can relate surface stress (σ) to a temperature gradient $(T_{avg} - T_{surf})$ where T_{avg} is the average temperature of the specimen and T_{surf} is the surface temperature of the specimen. It is necessary to employ physical constants of elastic modulus (E), thermal expansion coefficient (α), and Poisson's ratio (ν). The relationship is:

$$T_{avg} - T_{surf} = \frac{\sigma}{3\left(\frac{E\alpha}{1-\nu}\right)} \tag{7.4}$$

(For a full development of this relationship relating to thermal shock, see Section 7.9 in W. D. Kingery et al., *Introduction to Ceramics*).

By analogy of the value of $(m_{surf} - m_{avg})$ in Eq. 7.3 for $(T_{avg} - T_{surf})$ in Eq. 7.4 and rearranging, the result is:

$$\frac{\sigma_{surf}}{\left(\frac{E\varphi}{1-\nu}\right)} = \frac{1}{9}\left(\frac{Jw}{D}\right) \quad \textbf{(7.5)}$$

where φ is a shrinkage coefficient (cm/cm of shrinkage per unit change in moisture content) as an analogous linear change coefficient to the thermal expansion coefficient of Eq. 7.4.

The result of this analogy, Eq. 7.5, allows us to predict some things that we have understood about drying. For example, as the drying rate *(J)* increases, stress at the surface (σ) also increases for other factors held constant. As the body thickness *(w)* increases, surface stress (σ) increases. For a slower permeation rate *(D)*, surface stress (σ) increases.

For ceramic materials, we can assume that Poisson's ratio is constant at about 0.3. This suggests that the quantity (1 – ν) is about 0.7. If we choose a criterion that fracture will not occur if the following is approximately true:

$$\frac{\sigma}{E} \leq 0.05 \quad \textbf{(7.6)}$$

then we can define a maximum safe drying rate, J_{max}, by altering Eq. 7.5. The result is:

$$J_{max} \cong \frac{D}{w} \quad \textbf{(7.7)}$$

Equation 7.7 indicates that the maximum safe drying rate is directly proportional to the diffusion coefficient, the term describing the permeation rate for water during drying. In addition, D is also expected to be a function of concentration, and an exponential relationship between D and the concentration of moisture is reasonably expected. Therefore, the maximum safe drying rate when plotted versus concentration is not a constant (and a fixed maximum rate as is usually expected), but the maximum safe drying rate is likely an exponential function of the moisture content of the ceramic (Fig. 7.7).

This permits the practitioner to develop a family of plots of maximum safe drying rate versus half-thickness *(w)* as an aid in component design. To say the least,

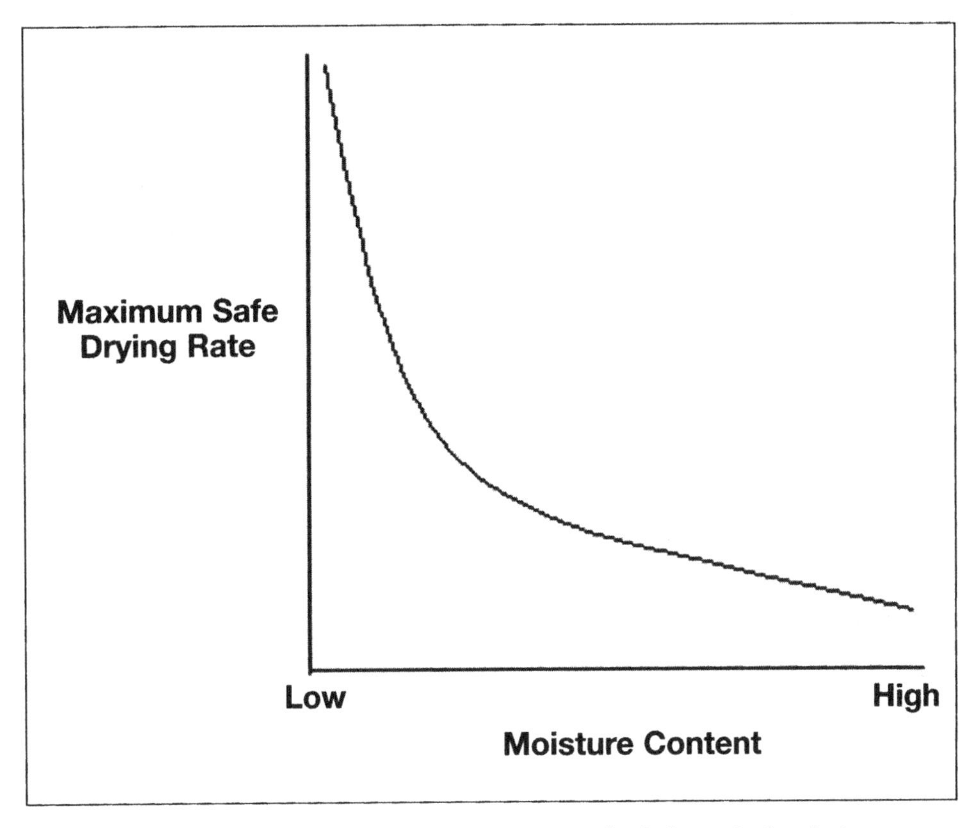

Figure 7.7: Form of the theoretical maximum safe drying rate for drying ceramics.

this is painstaking work. It is particularly difficult to observe the initiation of cracks in a drying ceramic component. However, use of time-lapse video recording equipment in humidity-controlled dryers with windows can minimize the effort expensed to generate the plots.

7.9 Problems

1. A common problem in extrusion is die skin crazing, a type of surface defect seen as surface cracking in extruded products. Discuss the reasons why there can be a higher potential for cracking near the surface of an extruded clay body and its effect on maximum safe drying rates.

2. A solid slip-cast rod-shaped object 6 in. long and 2 in. in diameter is laid on its edge during drying. It develops a crack in the center of the rod in a vertical direc-

tion to the setter plate in the dryer, and the crack does not extend to the edges of the object. Find the origin of the cracking and propose drying (process) solutions to eliminate the cracking.

3. Rod-shaped articles, made using a piston extrusion machine, are cut into 6-in. lengths. After drying, there is a predominance of cracking in the parts that were the last to exit the extruder. Consider moisture movement during extrusion (called dewatering), and propose a material solution so that you do not have to change the drying process parameters.

4. Brick are cut from an extrusion column of clay raw material, and after oven drying they are substantially cracked when placed in a flat set (large flat wise surface down) position. The cracks are primarily near the center of the brick. In a laboratory experiment, specimens from the same run of brick are dried and they achieve a bananalike shape after drying individual bricks on drying racks. What is the problem?

5. Explain why in slip casting of a whiteware ceramic, if the casting rate inexplicably decreases, you might also expect a decrease in drying rate or greater scrap in drying.

6. If you break a dried extruded clay ceramic part and see cracks on the cross section of the product like the ones shown below, what is the problem and what is a solution?

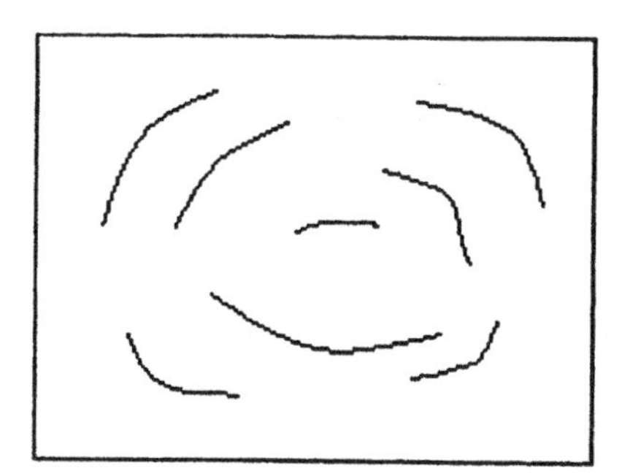

7. Pievsky et al. (cited in Additional Reading) provide experimentally determined moisture diffusion coefficients for different clays formed into bricks by dry pressing as follows:

Moisture content at optimal plasticity for dry pressing (%)	Diffusion coefficient $\times 10^{12}$ cm^2/s
2.0	3.0
1.0	4.5
0.5	7.5
0.2	14.9

Explain the fundamental reasons behind the observations given above.

8. During the course of drying experiments, Pievsky et al. measured the surface stress in ceramic plates made from different raw material mixtures at their optimal molding moisture contents:

Material	Moisture content for forming (%)	σ after 12 min of drying (N/mm^2)	σ after 18 min of drying (N/mm^2)
Kaolin mixture A	1.20	0.045	0.080
Kaolin mixture B	0.70	0.032	0.065
Kaolin with grog	0.35	0.0075	0.015

a. Develop an equation for diffusion coefficient versus moisture content using the data in Problem 8, and determine the diffusion coefficient for each mixture in the table above.

b. Use Eq. 7.7 to determine the relative maximum safe drying rate of each material using kaolin mixture A indexed to a maximum safe drying rate of unity.

9. Pievsky et al. provide data on surface stress for different drying rates for the kaolin with grog mixture described in Problem 9. Drying rate is described in terms of kilograms of water removed per unit surface area exposed per second of drying time. Data at 10°C are shown below:

Drying rate (J, $kg/m^2 s \times 10^4$)	σ (N/mm^2)
1.39	0.0002
2.78	0.0174
3.50	0.0380
4.50	0.0820

If you determine that cracking begins at a surface stress of 0.0345 N/mm^2 by very precise physical property testing, what is the maximum allowable drying rate for this material?

7.10 Additional Reading

W. D. Kingery et al., "Thermal and Compositional Stresses" (Chapter 16); in *Introduction to Ceramics*, 2nd ed. John Wiley and Sons, New York, 1976.

J. M. Pievsky, V. V. Grechina, G. D. Nazarenko, and A. I. Stepanova, "Intensification of Drying Process of Ceramic Building Materials due to Optimization of Raw Material Composition"; in *Drying of Solids: Recent International Developments*. John Wiley and Sons, New York, 1986. Pp. 227–230.

George W. Sherer, "Theory of Drying," *J. Am. Ceram. Soc.* 73 [1] 3–14 (1990).

CHAPTER 8

Advanced Drying Technologies

8.1 Introduction

There have been many attempts throughout history to increase the rate of drying of ceramic articles in order to shorten the cycle from 20–30+ h, as is typically employed in normal convection processes, to cycles requiring a minimum time. There are several obvious motivations for shortening drying time: the capital costs required for long tunnels, the expense of moving air around, and the cost of more kiln/dryer cars and track (or other conveyance such as trays and tray handling equipment). More recent motivating factors have been attempts to reduce energy requirements in processing by speeding up the drying process. Just-in-time manufacturing ideas are also becoming important in higher-value ceramic products.

One of the older techniques to increase drying rates is *rhythmic drying*, a technology developed in Europe in the late 1970s. Rhythmic drying uses intermittent air jets directed on the product to periodically remove the water layer present on the surface of the product during the initial stage of drying (Stage I). During dwell periods when the air jets are directed elsewhere or are turned off, water is transported back to the surface of the product. In this manner, the maximum surface stress is limited and surface cracking is minimized. A typical cycle might involve 10–15 s of air impingement with several minutes of dwell before the cycle is repeated. Rhythmic drying is practiced in some clay plants in the predrying stage, where the product is most sensitive to cracking. This latter practice may involve tunnel dryers with traveling or recripocating fans or it may involve recripocating fans placed on holding tracks simply to take advantage of the drying capacity of plant air.

Another process that receives periodic attention is *airless drying*. In airless drying, the product is placed in an enclosure capable of pressurization. The atmosphere in the container quickly achieves 100% relative humidity. Then the atmosphere in the container is heated in increments, causing accompanying pressurization; evaporation from the product is closely controlled by controlling the temperature in the enclosure. In effect, the atmosphere in the dryer becomes superheated steam. The term "airless drying" refers to the fact that the air flow of convection dryers is not required. When the product is dry, the atmosphere can be changed to low-humidity air, and the product is cooled to ambient temperature. While airless drying has advantages, it remains a batch process usable especially where waste heat for normal convection drying is unavailable.

Processes using electromagnetic radiation to affect drying have been known for a number of years. *Radio frequency* was the first process employing electromagnetic radiation generally applied to drying of technical ceramics (~1970). Radio frequency drying is also called *dielectric drying* and *RF drying*. While RF techniques have been known since the 1940s (and earlier), the application to ceramic drying was limited by the cost of generators and the need for an application where RF drying would be of great advantage. As explained below, this advantage was found in electrical insulators and catalytic reactor materials.

Microwave (MW) drying was applied to ceramics in the late 1990s. While RF drying uses a frequency of about 25 kHz, microwave dryers use radiation at a frequency as high as 2.45 GHz. This implies that the longer wavelength of RF radiation would penetrate the drying ceramic to a greater degree, resulting in faster drying than in the case of MW drying. The difference in drying time between RF and MW drying has not been found to be significant in some cases, and MW dryers offer some advantages for ceramic processing.

Duplex drying techniques — those involving a combination of techniques — will likely appear in the future. Because the efficiency of both RF and MW dryers decreases as the moisture content decreases (in Stage II of drying), these dryers may employ hot air to more efficiently accomplish the last portion of drying. In the case of turned ceramic products (those formed by jiggering processes after a blank has been formed), joule heating may be used on the extruded blank to reduce the moisture content of the blank slightly until the object is sufficiently strong to be handled. The formed objects are then microwave dried.

Other applications for microwave drying are being developed for tableware, tile, sanitaryware, and refractories. The application of microwave drying in a slip casting operation for sanitaryware has been shown to dramatically increase throughput by drying both the part and the mold while decreasing. With steel-plant refractories, such as molten metal contact nozzles or pouring shrouds, MW heating is being used prior to immersion of the ceramic in molten steel. As more commercial microwave systems become available, the utilization of this technique will likely expand.

PART A: MICROWAVE DRYING

8.2 Physics of Microwave Drying

Microwave energy is electromagnetic radiation that falls in the frequency range 0.3–300 GHz. Radio frequencies are adjacent to the microwave and radar section of the electromagnetic spectrum, but with a longer wavelengths (lower frequency) than microwaves. Electromagnetic waves act on materials that have a specific susceptibility or absorption of energy at these frequencies. Polar molecules, like water, tend to be susceptible to energy at microwave or radio frequencies. The charged portions of these molecules couple with and move with respect to the oscillating electric field of the electromagnetic waves over specific frequency ranges that are unique to the structure and chemical makeup of the molecule. This movement of charged regions of the dipole relative to each other is realized as molecular vibrations that result in kinetic heating. This excitation can also result in rotational displacements within the polar molecule that can also result in heating. In addition to these short-range influences of the absorption of microwave radiation, long-range conduction processes that involve the transport of charge can result.

For the drying of traditional ceramics, microwave drying is theoretically more effective than drying with radio frequencies due to the fact that water couples with electromagnetic radiation at microwave frequencies more readily than at radio frequencies. For MW drying, the specific loss factor of water with respect to microwave frequencies is used to control the rate and quantity of water removed from a ceramic article. For water and other materials, the susceptibility to energy at a particular frequency is a function of temperature and phase composition. For example, liquid water is the only phase of water that absorbs microwave energy while solid water (ice) and steam are transparent to microwave frequencies. Water is susceptible to MW energy over a wide frequency range while for RF, water is only reasonably susceptible if it has a high dissolved salt content. For water, the maximum MW absorption occurs at about 18 GHz, although the most common frequency for commercial microwave systems is 2.45 GHz.

The driving force for microwave drying is similar to conventional hot air drying, but the notable physical difference is that there is internal heat generation with MW heating. The water in the formed ceramic absorbs the microwave radiation through its cross section (at least partially), and this absorption generates heat through the bulk of the part that can be used for evaporation. For conventional hot air drying, in contrast to microwave drying, the heat for evaporation is transferred to the surface of the part by convection and then through the bulk of the part by conduction.

The following is an idealized description of the drying process with microwave drying:

1. In the first period of drying, heating occurs within the ceramic and the temperature increases to the boiling point of the liquid phase due to absorption of microwave radiation.
2. In the second period of drying, the pressure within the part begins to increase to its maximum value due to the resistance to transport of the pore water phase to the water vapor that is being developed from evaporation.
3. At constant power application, the next phase of drying is the constant rate period in which the rate of vapor transport is controlled by the power absorption of the water's interaction with the pore network's resistance to vapor transport.
4. The last phase of the drying process is the falling rate period, in which the reduction of the moisture level of the part decreases the amount of power that the part can absorb, which in turn reduces the driving force for evaporation and transport. The body temperature of the drying part can increase substantially if the material is susceptible to microwave radiation (as is the case with most ceramic materials).

Other factors, such as the hygroscopic nature of the material to be dried, the purity of the liquid phase (latent heat of evaporation), and changes to the vapor and heat transport properties of the part, can all significantly influence the process.

The power absorbed by the part in a microwave dryer can be calculated according to the following equation:

$$P_{\mathrm{A}} = 2\pi f \varepsilon_0 \varepsilon''_{\mathrm{eff}} |E|^2 \tag{8.1}$$

where: P_{A} = power absorbed per unit volume.

f = frequency of the field.

ε_0 = permittivity of free space.

$\varepsilon''_{\mathrm{eff}}$ = relative loss factor that includes all of the active loss mechanisms (vibration, rotation, etc.).

E = root mean square value of the internal electrical field per unit volume.

Further, the depth of penetration, which is defined as the depth from the surface at which the power drops to e^{-1} of the incident strength, can be described according to the following equation:

$$D_{\mathrm{P}} = \frac{\lambda_0}{2\pi(2\varepsilon')^{1/2}} \left\{ \left[1 + \left(\frac{\varepsilon''_{\mathrm{eff}}}{\varepsilon'} \right)^2 \right]^{1/2} - 1 \right\}^{1/2} \tag{8.2}$$

where λ_o is the free space wavelength of the microwave radiation and ε' is the relative dielectric constant. The effective loss factor ε''_{eff} is a frequency-dependent term that describes the material's ability to absorb electromagnetic radiation.

The loss factor is related to the relative dielectric constant by:

$$\varepsilon''_{eff} = \varepsilon' \tan\delta \quad (8.3)$$

where $\tan \delta$ is the loss tangent of the δ-phase angle between two currents. As the value of δ increases, the greater the time lag between the maximum of the electric field and the maximum polarization of the material becomes.

Absorption and reflection of the incident microwave radiation can influence the actual amount of microwave power that is converted into heat within the drying part. The influence of absorption and reflection of the radiation at the surface can be taken into account according to the following equations:

$$P_0 = \frac{P_A}{\left(1-\Gamma^2\right)e^{-2\alpha y}} \quad (8.4)$$

where P_0 is the power applied to the surface of the part to be dried, y is the distance of the surface from the source, and Γ is defined as:

$$\Gamma = \frac{\sqrt{\varepsilon'} - 1}{\sqrt{\varepsilon'} + 1} \quad (8.5)$$

and the attenuation constant α is defined as:

$$\alpha = \frac{2\pi f}{c}\sqrt{\frac{\varepsilon'}{2}\left(\sqrt{1}\right) + \tan^2\delta - 1} \quad (8.6)$$

and c is the heat capacity of the system at constant pressure. Due to the influence of absorption and reflections in the chamber, the value of the power applied to the surface (P_0) is usually higher than the absorbed power (P_A).

8.3 Advantages of Microwave Drying

Due to the high frequency and short wavelength of microwave radiation, there are several advantages to MW drying, such as greater energy efficiency and more uniform drying. With conventional hot air drying, the heat is transferred to the surface of the part by convection, and then the heat is transferred to the interior by conduction. As the moist part is heated sufficiently, water evaporates at the surface and the heat is progressively transferred deeper into the part to affect evaporation and transport of water out of the piece. A consequence of the water removal is that the part shrinks as the water is removed from between the grains. The magnitude of the shrinkage is dependent on the particle size distribution and particle packing of the part as well as the composition of the part. The distinct disadvantage of this scheme is that the pore structure shrinks as the material dries from the surface inward, so the remaining water vapor must be removed through an increasingly restricted path. If the restriction is too great or the pressure of the water vapor is too high, cracking or explosion of the part can result.

Unlike convective heating, which acts only on the surface of the part, microwave drying is more uniform over the part's cross section due to the penetration depth of the microwave radiation. This means that the water is evaporated more uniformly instead of just at the surface, and there is less change to the pore network at the beginning of drying, resulting in easier transport of the water vapor out of the piece. In other words, for traditional convective drying, the shrinkage takes place from the surface and proceeds inward, while for microwave drying the shrinkage begins at the center of the part and progresses to the surface. With microwave drying, typically the center of the part achieves the highest temperature and dries first due to resonance effects from the microwave energy, impurity content of the water phase, and water distribution in the piece.

Usually, for microwaves the depth of penetration is also many times greater than the thickness of the part. This implies that there is no penalty to the rate of drying from stacking parts as there is in convective drying processes. For convective processes, the outer pieces get heated preferentially by the impingement of the hot air flow while inner pieces do not receive direct air contact and must rely more heavily on conduction for heat to evaporate water.

It is important to remember that with microwave drying it is still possible to crack or explode parts due to excessive application of microwave power. If the rate of steam generation within the part exceeds the ability of the part to transport the vapor to its surface, an unacceptable pressure buildup can result that exceeds the strength of the part and cracking or explosion can ensue (traditionally called *shelling* in ceramic processing). An additional advantage of microwave drying is that the heating is instantaneous due to its molecular nature and it can be rapidly switched on and off. In fact, varying time on and off (and power input) is the normal method to control MW dryers.

8.4 Microwave Drying Systems

Microwave energy is produced using magnetrons. There are several systems for controlling and directing the microwave energy to the part to be dried. A magnetron consists of an electronic valve that is under a high vacuum and is made of a hollow copper anode with a resonant microwave structure and an electron-emitting cathode at its center. The microwave energy that is produced is transmitted to the applicator by means of a rectangular, precisely machined metal tube known as a waveguide. Metals are used in the waveguide since, in general, they reflect microwave radiation.

Since the MW energy is produced whether or not it couples with the part to be dried, there are provisions to absorb excess energy, such as dummy loads and isolators. The first type of microwave system is the *multimode system* in which the part to be dried is situated inside a metal cavity. The incident radiation from the magnetron is introduced via the waveguide and is reflected by the metal walls of the chamber, and standing waves with nodes and antinodes develop. To distribute this energy evenly, a mode stirrer is used, which consists of a metal paddle that is rotated and changes the standing wave patterns. For a similar effect, the part can be rotated so that it moves through the standing wave pattern and distributes the incident energy more uniformly. Multimode systems are employed in laboratory batch MW dryers.

A second type of microwave system is the *single mode resonant cavity* in which an incident microwave radiation from the magnetron undergoes multiple reflections between preferred directions within the metal chamber. The overlapping of the incident and reflected waves gives rise to a standing wave pattern. This type of system allows the part to be placed in the area of maximum field strength, and it allows for higher power densities to be applied than in the previous system.

The last major type of system is the *traveling wave applicator* in which a rectangular waveguide is used and the part to be dried is introduced continuously, as in sheet form. This would be applicable for a tape casting system, for example.

Microwave dryers can employ a number of individual magnetron generators. Advantages of this approach include energy distribution in the dryer and lower replacement costs of lower-capacity units. Magnetrons employ cooling air in their operation, and this heated air can be used to assist in drying, particularly in the final stage of drying.

8.5 A Practical Example of Microwave Drying

Green (plastic or undried) brick of the following characteristics were MW dried in a series of experiments:

- Hand molded modular brick.
- Extruded paving brick (solid brick). Brick were "triple size" (thin) units 8.5 × 4.6 × 4 in. and normal size 8.5 × 4 × 3 in. The triple size units represented a worst case drying situation.

Brick were dried in a laboratory batch-type oven rated at 8.7 kW. Brick were set on a ceramic pedestal (Fig. 8.1) with microwave energy supplied from six mag-

Figure 8.1: Batch microwave dryer with a single brick in place.

netrons located in the roof of the dryer. The dryer interior was stainless steel to ensure reflection of radiation from the magnetrons. The specimen support allowed for continuous monitoring of weight. A thermocouple was inserted into the center of one brick for each experiment, and the internal temperature was used as a control variable. The power input from the MW generators was altered as necessary to maintain a specified heating schedule (based on the brick internal temperature). Brick were evaluated based on visual observation of defects.

The experimental strategy was to conduct drying trials of the hand-molded bricks first. Then, the extruded pavers were dried as a most difficult case (based on the large size). The results are shown in Table 8.1.

In general, the bricks were found to heat and generate steam in about 30 min of microwave exposure at approximately 17% of full power. Excessive rise in internal temperature was found to result in cracks in all types of brick (Fig. 8.2). Hand molded bricks exhibited a central shrinkage crack with excessive initial heating rates. As there was no ability to observe the specimens during drying, a practice was developed to stop drying periodically and examine the specimens. Excessive initial heating of extruded brick was found to result in edge cracks. In all cases, the reaction to cracking was to reduce heating rates in a trial-and-error process.

Table 8.1: MW experiments with brick

Trial	Type	Schedule (temp. of brick center)	Starting moisture (%)	Ending moisture (%)	Comments
1	Hand-molded modular	81→175°F in 1 h (steaming) 175→248°F in 1 h 36 min Total 2 h 36 min	23.6	18.7	4.9% residual moisture.
2	Hand-molded modular	81→175°F in 30 min 175→248°F in 40 min 248°F hold 30 min 248→626°F 1 h 55 min Total 3 h 35 min	23.6	~0	Considered successful. No visual defects. Decided to increase power in first period in next trial.
3	Hand-molded modular	81→212°F in 30 min 212°F hold 60 min 212+°F 30 min	23.6	NA	Shrinkage cracks noted (60% of power applied in first period in Trial 3 vs. 23% in Trials 1 and 2. Trial suspended.
4	Extruded solid 8.5 × 4.6 × 4 in.	81→204°F in 40 min	16.75	NA	Shrinkage cracks in 40 min (@2.1% H_2O removed, 0.9% linear shrinkage). Trial suspended.
5	Extruded solid 8.5 × 4.6 × 4 in.	81→212°F in 60 min	16.75	NA	Shrinkage cracks in 60 min (@2.7% H_2O removed, 1.1% linear shrinkage). Improvement over Trial 4. Trial suspended.
6	Extruded solid 8.5 × 4.6 × 4 in.	81→140°F in 30 min 140→160°F in 30 min 160°F hold 48 min	16.75	9.35	Shelling @ 7.4% moisture removed with 2.1% linear shrinkage.
7	Extruded solid 8.5 × 4.6 × 4 in. and extruded solid 8.5 × 4 × 3 in.	81→140°F in 60 min 140°F hold for 60 min 140→175°F in 60 min 175→212°F in 60 min 212°F + air hold 60 min Total 5 h	16.75	0.7 for 8.5 × 4.6 0.8 for 8.5 × 4	Considered successful.

Figure 8.2: Cracking in hand-molded bricks due to excessive initial heating rates.

Extruded brick exhibited shelling or steam spallation after the initial drying period (>1 h of drying time) if continued heating rates were excessive (Fig. 8.3). This was seen as a loss of corners in solid bricks or loss of facial sections in three-hole bricks. As with cracking, the response was a reduction in heating rates.

The batch dryer exhibited condensation on the interior walls of the dryer. This ensured a near 100% relative humidity condition within the drying chamber during the initial stage of drying. As air movement was not employed in the MW trials, this environment was maintained throughout drying. This is counter to practice in convection drying. In MW drying, the local atmosphere around the product is apparently heated sufficiently to allow drying even though condensation occurs on the walls of the dryer (indicating 100% relative humidity in the air adjacent to the walls). A commercial MW dryer would likely allow for continuous moisture removal from the dryer chamber during the final stages of drying.

The successful drying trial for hand-molded brick was Trial 2 (drying to 0% moisture in 3 h, 35 min); for extruded paving brick, Trial 7 (drying to <1% moisture in 5 h).

Figure 8.3: Shelling in extruded bricks due to excessive heating rates in the final stage of MW drying.

PART B: RADIO FREQUENCY DRYING

8.6 Brief Description of Radio Frequency Drying

Radio frequency drying follows the same physical laws presented in Part A with respect to radiation transfer into the object to be dried. Like MW drying, RF drying can be applied to dielectric (nonconducting) materials like ceramics. RF drying was applied to drying of clay-based electrical insulators in the 1960s. These products are turned ceramics. Subsequently, extruded thin-wall catalytic reactor parts were dried by RF because it was advantageous to immediately dry the formed parts, which have limited green strength.

RF dryers have an internal "antenna" that extends across two sides of the dryer chamber. In some commercial RF dryers of the continuous tunnel type, the antennas

Figure 8.4: Commercial RF dryer processing ceramics.

may be arranged on the top (over the traveling load) and bottom (under the traveling load). A commercial RF dryer is shown in Fig. 8.4.

8.7 A Practical Example of Radio Frequency Drying

The hand-molded brick described in Section 8.24 was dried using a batch RF dryer (27 kHz at 1.5 kW) and a continuous RF dryer (27 kHz at 25 kW). The experiments in the batch RF dryer proved that drying was feasible, and the continuous RF dryer was used in later experiments (Table 8.2). The conclusions from the studies were:

- Power input must be restricted during the initial period of drying to avoid formation of shrinkage cracks. Power was limited by maximizing the distance between the transmitting electrodes located on two sides of the brick.
- Studies apparently employed flat set brick, and it was necessary to allow for movement using parting sand to avoid cracking.
- Completely dry and crack-free brick were produced using RF in 3 h using a two-stage process (75 min of low power followed by 30 min of high power in Trial 25).
- A duplex process of 75 min of RF followed by convection heating for 30 min (presumably using waste heat in a commercial installation) produced dry and crack-free brick.

Table 8.2: RF drying trials with hand-molded brick

Trial	Time in Zone 1 (plate spacing)	Time in Zone 2 (plate spacing)	Convection oven step	Total drying time	% H_2O removed
24	180 (370 mm)			3 h	81.9
25	75 (370 mm)	30 (260 mm)		3 h	100.0
26	75 (370 mm)	30 (300 mm)		1 h 45 min	96.7
27	75 (370 mm)		30 min @ 430°F	1 h 45 min	100.0
28	75 (370 mm)	15 (300 mm)		1 h 30 min	44.5
29	75 (370 mm)	15 (300 mm)		1 h 30 min	50.1

8.8 Costs of Microwave and Radio Frequency Drying

The variable cost of MW drying (electrical costs at $0.05/kWh) is calculated at about $0.05/brick; a similar cost was estimated in the RF drying study. This implies that MW or RF drying would be used in brick plants for higher-value products (such as special shapes) or other products where manufacturing time is important. Chamber MW dryers are available with the following specifications:

Design:	Chamber MW dryer with automatic doors
Transport:	Roller conveyor
Dimensions of chamber:	78 in. L × 47 in. W × 47 in. H
Useful width:	39 in.
Useful height:	27 in.
Total length with conveyor:	19.7 ft
Number of controlled magnetrons:	16
Connected load:	35 kVA
Voltage:	230/400 V, 3 phase

Continuous MW dryers are available, and they are used in dinnerpart drying. In the future, the application of RF or microwave dryers will continue where speed of manufacturing is of importance. As product value increases, the chances for application of MW and RF drying becomes economically attractive.

PART C: AIRLESS DRYING

8.9 Introduction

Airless drying is a new proprietary technique that offers the potential to greatly reduce drying times and energy costs while improving product quality. This procedure utilizes superheated steam to enhance heat transfer and control the evaporation rate of water from the part. It was originally designed as a batch system, but recently continuous airless dryers have been described. This technology is patented by Heat-Win Ltd in the United Kingdom and is also licensed through Ceram Research.

Instead of heating a continuously flowing atmosphere, as with traditional hot air drying, airless drying relies on a strictly closed atmosphere where a saturated steam atmosphere is used as the heat transfer medium. In traditional drying, the part is heated as evaporation takes place, which results in temperature, moisture, and permeability gradients throughout the piece. These can result in reduced drying rates and cracking or other defects if drying is allowed to proceed too quickly. With the airless drying technique, evaporation of water is severely limited until the entire part has been heated to the desired drying temperature, which results in faster evaporation rates and more homogenous drying. Since the part is of uniform temperature, drying occurs over the whole volume of the piece and not just at the surface as in traditional hot air drying. This results in a lessening of moisture and permeability gradients, which leads to fewer defects and increased drying rates. A schematic of an airless dryer is depicted in Fig. 8.5. As can be seen, the dryer works with a closed, recirculated atmosphere. When necessary during the drying phase, steam is vented only to control the pressure in the dryer and to control the drying rate in the second stage of drying. This closed atmosphere offers the immediate cost saving of not continually heating incoming ambient air to dry the part.

The first stage of drying is known as the priming stage (see Stubbing and Ford in Additional Reading). After the part is loaded into the airless dryer, it is sealed and heated indirectly by a gas burner. The atmosphere that is contained in the dryer is recirculated and heated continuously to bring the temperature of the dryer system to approximately 100°C. A typical profile of an airless drying cycle is depicted in Fig. 8.6. As the charge and dryer atmosphere are heated, some water is evaporated from the surface of the green part and the steam content of the atmosphere increases until the dryer atmosphere achieves saturation for the particular drying temperature. Air and steam are vented from the dryer to maintain atmospheric pressure in the dryer. When the dryer atmosphere is saturated, the priming stage is complete and the actual drying of the part is allowed to proceed.

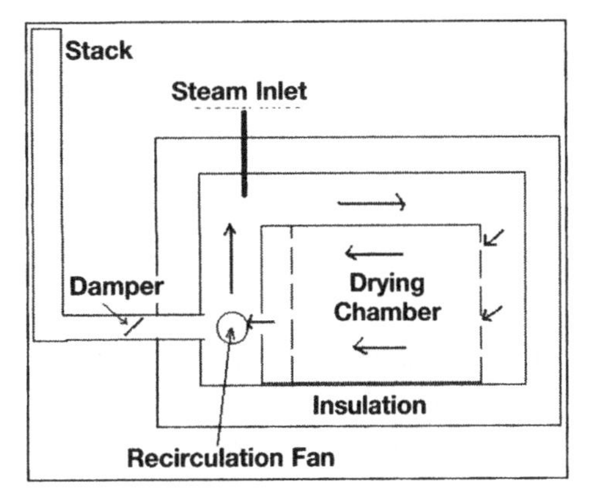

Figure 8.5: Schematic of airless dryer.

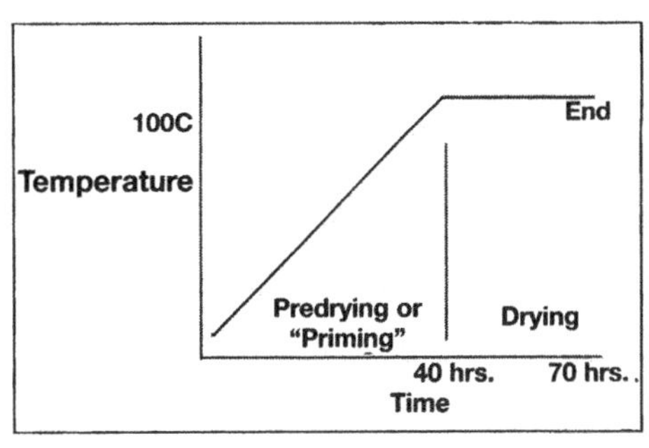

Figure 8.6 Typical airless drying cycle.

During the drying phase, heating is continued until the final drying temperature is achieved (Stubbings and Ford). The saturated steam atmosphere of the dryer serves to transfer heat into the part more efficiently than moving hot air in a conventional dryer. The heat from the steam is used to evaporate water from the part. As this water is evaporated, the pressure in the dryer increases, which is again vented to maintain atmospheric pressure. As the supersaturated atmosphere is vented from the dryer, additional evaporation occurs from the part, and as such, the rate of steam release controls the drying rate. Since the temperature of the part cannot exceed 100°C while drying is taking place, an increase in temperature of the part indicates that all of the moisture has been evaporated and the drying stage is complete. This increase in temperature is also indicated in Fig. 8.6.

This airless drying technique shares many similarities with microwave drying in that the drying is much more homogenous than traditional hot air drying. Since the part is brought to the drying temperature before the majority of the drying is allowed to take place, the drying proceeds through the volume of the part. Conversely, for traditional hot air drying, hot air impinges on the surface of the part that causes evaporation at the surface. This evaporation at the surface typically results in shrinkage and restriction of the pore phase, which hinders further transport of water from deeper in the part. The result of this type of drying is that the part dries from the surface into the center of the part and as the moisture is evaporated from deeper in the part, it has an increasingly difficult path to reach the surface of the part. More volumetric heating, as in the airless drying technique, where evaporation takes place across the cross section of part, results in the part drying from the interior to the surface. This condition means that the path of the evaporated moisture to exit to the surface is not restricted during drying and faster evaporation rates can be safely achieved.

Despite the advantages of this method, it is not without hazards. As with any drying process, cracking or spalling can occur if the evaporation rate is too high and the pressure of the evaporated moisture in the pore phase exceeds the strength of the piece. Similarly, if the initial heatup is too fast, cracking can result from the surface drying too soon and reducing the permeability of the part, which inhibits further transport of water to the surface.

8.10 Semicontinuous Systems

While the previously described system is best suited for batch systems, Heat-Win Ltd has recently described a semicontinuous system. In this system, the part to be dried is conducted via an open duct upward into the drying chamber. Once in the drying chamber, the part is dried much like in the previously described drying system. After the cycle is complete the part is conducted down and out of the drying chamber through a similar duct. Venting the saturated steam atmosphere just above the duct opening effects a seal. This gives a steam curtain effect that maintains the saturation of the atmosphere in the drying chamber.

8.11 Advantages of Airless Drying

There are several advantages associated with the airless drying technique as opposed to traditional hot air drying (see Small in Additional Reading). Airless drying provides increased flexibility through reduced drying time and improved quality. In industrial trials to date, airless drying has offered up to a 95% reduction in energy consumption as compared to a traditional hot air dryer. The reduced energy consumption is due to the fact that there is no need to heat a continuous supply of fresh, dry air and that the heat loss through the flue of the dryer is substantially reduced. Only saturated air with a temperature of less than 100°C is vented, which means that the majority of the input energy has been consumed in evaporation and heating the part. Since higher evaporation rates can be safely used with airless drying, reduced drying time is possible. For a case of drying cast parts, a reduction of the drying time from 18 to 2 h has been reported (Small). As an added bonus, distilled water can be easily recovered from the system, which can then be used in another phase of the process. A further cost advantage is that since this water is clean, it does not require any treatment or special handling.

PART D: FAST DRYING USING CONVECTION

8.12 Introduction

Fast drying developments in Europe were reviewed by Karsten Junge, Director of the Institut für Ziegelforshung, in the July 2000 issue of *Ziegelindustrie International.* The original fast drying research was based on the development of large, thin, story-high units called *plank bricks.* These units included a thin web and many cores to reduce their weight. Although this pioneering work of the late Carl Otto Pels Leusden on plank brick did not provide a commercial product, it gave the world the drying technology that we know today as *fast drying.*

What Dr. Leusden discovered was that stress-free drying is possible if the units are sufficiently cored with a small enough web and face shell thickness of clay. This is a very important contribution, and it results in the fact that ability to remove water becomes the rate-limiting step, rather than the rate of water permeation through drying clay. This is fundamentally a new drying strategy.

According to Dr. Junge, fast drying employs "the most intensive air mixture possible and high air rates impinging on the green brick surface as the essential condition for intensive and uniform drying." Air flow must be directed through the cores in the brick to increase the drying surface area and to minimize the path length for moisture permeation in the green body. Further, air flow must be directed perpendicularly toward cored units with units spaced apart the same distance as the web thickness. This avoids eddy currents of drying air, "shadows" in air movement, and other nonuniform air flow conditions.

The rewards of fast drying are drying highly cored brick in 4.5 h instead of 20+ h. Even if North American facing bricks, classed as solids under ASTM C-216 (25% core area maximum), dried in 15 h instead of 20 h, the gain is 25% greater drying capacity in the same size dryer. With hollow units classed under C-652, even greater drying productivity could be expected.

8.13 Fast Drying Plants

Early plants arranged cored brick that were flat set on expanded metal shelves with air impingement from above. Designs evolved to use stacked shelves with the product edge-set using air impingement from a horizontal direction. The stacked shelves are held in a frame called a *casing pallet,* and two casing pallets are arranged on a single dryer car with a freeboard space (of the approximate size of a manway) between casing pallets. This freeboard space forms the plenum for drying air in a plant owned by Rapis in Schwabmünchen, Germany (Table 8.3, Fig. 8.7; plant and

Figure 8.7: Fast dryer entrance at Rapis, showing the casing pallet design.

Figure 8.8: Casing pallet and transfer car at the Rapis plant.

Table 8.3: Plant Information

Feature	Typical Value	Comment
Plant Layout (Operating Side)	30 30 meters	Approx. 9680 square feet
Plant capacity	190 metric tons/day at about 17 kg/unit	Approx. 172 short tons/day
Plant personnel	2 operators/shift with 12 shifts/week	
Dryer	1 forward and one return track at 12 cars each	Cars travel in a "U" shaped path entering and exiting in close proximity.
Dryer length and width	42.80 meters × 9.00 meters	140.4 feet × 29.5 feet
Heat consumption	4200 kJ/kg water	1806 BTU/lb. water
Power consumption	12 kWh/ton	
Drying time	4.6 hours average	
Kiln Length	89.7 meters with 23 cars	294 feet
Firing time	14 hours	

dryer/kiln by Lingl Corporation). In this plant, hot air enters the dryer from ducts in the roof of the dryer.

The casing pallet is shown to contain ten shelves with six large, highly cored structural brick set one deep per tray (Fig. 8.8). The plant has the capability to set two deep per tray. The ventilating equipment injects drying air into the freeboard space between casing pallets, and air travels through and around brick arranged on the shelves.

To make the case for rapid drying even more solid, the raw material used at Rapis would normally be considered somewhat difficult to dry. With 80% –200 mesh and 24% finer than about 2 μm (distribution modulus n of 0.20), high drying shrinkage would be expected in a product made by soft extrusion. Rapis uses a small amount of sawdust and other nonplastics in its body. Yet brick exiting the dryer clearly exhibited no face cracks and only a few web cracks. Further refinements have been made in the rapid dryer design since Rapis was built. The latest fast drying plant has airflow injected through ducts entering the side of the dryer.

8.14 Problems

1. Calculate the depth of penetration of electromagnetic radiation in the ceramics given below using equations 8.2 and 8.3:

Problem	Frequency	ε'	$\tan\delta$
(a)	RF (27kHz)	6.60	0.0009
(b)	MW (2.45GHz)	6.60	0.0003

Hint: The relationship between frequency (ν) and wavelength (λ) is:

$$\nu \lambda = c$$

where c is the speed of light.

2. Repeat the calculation of Problem 1 using the dielectric constant (ε') of water of 81 (at 20°C) and a value of $\tan\delta$ of 1.0000. Compare the depth of penetration when water is present to the case when water is absent (Problem 1).

3. Using Eq. 8.1, argue that microwave radiation has superior power absorption as compared to RF radiation (while RF radiation has superior penetration over MW radiation as shown in Problems 1 and 2).

4. Given molded brick at 23.6% moisture (WB) used in MW drying Trial 1 (Table 8.1). The 8.7 kW dryer was operated at 20% of rated power for the 2 h, 36 min drying period (leaving 4.9% residual moisture). What was the efficiency of the drying process (energy input in kWh/theoretical required) to evaporate 18.7% of the water present and raise the temperature of the brick to 248°F?

5. Repeat the calculation requested in Problem 4 for the brick in Trial 7 of Table 8.1.

6. A common defect in MW and RF drying of hand-molded brick is generation of face cracks during the initial stage of drying (as steam evolution begins). Practical experience shows that these cracks are generated at a time when the weight loss of the overall brick unit is less than 1%. Discuss strategies to eliminate this cracking from the aspect of the particle size distribution of the clay.

7. Discuss ways to reduce the tendency of extruded brick to shell during microwave drying (see Trial 6 in Table 8.1).

8. What influence does the thickness of the piece (brick) have on the occurrence of shelling such as that seen in Trial 6 of Table 8.1?

9. Given the fact that the metal surfaces of a MW dryer reflect radiation while the ceramic load absorbs MW radiation, sketch a design of a dryer intended to dry 100 bricks at a time, achieving the highest possible efficiency. Show the setting pattern (stacking pattern) for products inside of the dryer.

10. Explain why the efficiency of MW or RF drying decreases rapidly during the last stage of drying and find methods to complete drying in an efficient manner.

8.15 Additional Reading

I. Chablinsky and E. Eves, "Application of Microwave Energy in Drying, Calcining and Firing of Ceramics," *Ceram. Eng. Sci. Proc.* **6** [11-12] 1412–1427 (1985).

D. Clark and D. Folz, "Developments in Microwave Processing Technologies," *Ceram. Eng. Sci. Proc.* **18** [4] 531–541 (1997)

D. Clark, W. Sutton, and D. Lewis, "Microwave Processing of Materials"; pp. 61–96 in *Microwaves: Theory and Application in Materials Processing IV.* American Ceramic Society, Westerville, Ohio, 1997.

L. Dissado and R. Hill, "What Form of Motion Gives Rise to Dielectric Response at Microwave Frequency"; pp. 99–106 in *Microwaves: Theory and Application in Materials Processing IV.* American Ceramic Society, Westerville, Ohio, 1997.

D. Earl, "The Feasibility of Microwave Drying Ceramic Tile," *Ceram. Ind.* **146** [11] 30–34 (1996).

W. Evans, "Rapid Casting Trials in a Micro-wave Drying Oven", *Trans. J. Brit. Ceram. Soc.* **72** [8] 365–369 (1973).

Heat-Win Limited Web page: <www.dryers-airless.mcmail.com>.

W. Hendrix and T. Martin, "Microwave Drying of Electrical Porcelain: A Feasibility Study," *Ceram. Eng. Sci. Proc.* **14** [1-2] 69–76 (1993).

M. Kennedy "Commercialization of Microwave Processes"; pp. 43–54 in *Microwaves: Theory and Application in Material Processing III.* American Ceramic Society, Westerville, Ohio, 1995.

Ö. Perhahl, "Application and Operation Results of the Microwave Technique in Drying of Steel Plant Ceramics, Especially Snorkels of RH-Degassers," *Interceram* **40** [2] 75–76 (1991).

E. Schlünder, "Microwave Drying of Ceramic Spheres and Cylinders," *Chem. Eng. Res. Design Part A* **71** [11] 622–628 (1993).

H. Segerer, "Producing Technical Ceramics by Microwave Drying," *Bull. Am. Ceram. Soc.* **77** [3] 64–66 (1998).

L. Sheppard, "Manufacturing Ceramics with Microwaves: The Potential for Economical Production," *Bull. Am. Ceram. Soc.* **67** [10] 1656–1661 (1988).

G. J. Small, "Improved Business Performance through Advanced Technology," *Interceram* **44** [3] 180–185 (1995).

C. Strumillo and T. Kudra, *Drying: Principles, Applications and Design.* Gordon and Breach Science Publishers, New York, 1986. Pp. 376–406.

T. J. Stubbing and R. W. Ford, "Airless Drying of Ceramic Products," *Brit. Ceram. Trans. J.* **91** [3] 100–102 (1992).

M. Szczepanski and J. Szczepanski, "A Microwave Effect: Molecular Level Microwave Study of Water Vapor"; pp. 107–114 in *Microwaves: Theory and Application in Materials Processing IV.* American Ceramic Society, Westerville, Ohio, 1997.

Villeneuve, R. Girard, M. Giroux, J. Goyette, J. Kendall, and J.-M. Arnaud, "Microwave Characterization of a Ceramic Green Body for an Industrial Drying Process," *Interceram* **41** [3] 143–147 (1992).

[illegible] "[illegible]," [illegible] (1998).

[illegible] [illegible] (1994). pp. [illegible]

[illegible] (19[illegible]).

[illegible] "[illegible]," [illegible] Materials Research Society Symposium Proceedings, [illegible] (1992).

CHAPTER 9

Laboratory Exercises

9.1 Introduction

Laboratory exercises are designed for a one-term undergraduate experience in drying and firing of traditional ceramics. The philosophy of these experiments is that they employ the most fundamental types of procedures and the least expensive equipment possible. In this sense, they are analogous to ASTM tests on traditional ceramic products.

These experiments are designed to use one raw material throughout a term. For example, a ground brick mix from a local manufacturer could be used. It is also possible to use only a few of the experiments and to substitute more advanced methods where appropriate. For example, fine particle sizes could be determined using a laser light scattering technique instead of the traditional sedimentation technique.

The topics covered in the laboratory exercises are:

1. Particle size: screen analysis for dryer control.
2. Particle size: sedimentation analysis for dryer control.
3. LOI and moisture absorption.
4. Extrusion of clay bodies: effect of moisture content.
5. Stages of drying.
6. Permeability in drying.
7. Mix pelletizing/pressing.
8. Thermal expansion and gradient furnace.
9. Firing experiment, properties of extruded clay bodies.
10. Firing experiment, properties of pressed clay bodies.
11. Mercury porosimetry and SEM.

The equipment needed for each experiment is given in Table 9.1.

9.2 Suggested Laboratory Safety Rules

The following safety rules are typical of laboratories where clay is being processed. Local laboratories where experiments are conducted bear the responsibility

Table 9.1. Equipment list for laboratory exercises

Lab	Equipment
All	Top-loading balances
1. Screen analysis	Ro-tap® test sieve shaker and screens
	Oven
2. Sedimentation analysis	High-intensity mixers
	Clay hygrometers
	Sedimentation column (1000 mL graduated cylinders)
3. LOI and MA	Ignition furnace
	Humidity chamber
	Porcelain or alumina crucibles
4. Extrusion	High-intensity mixer
	Lab extruder
5. Stages of drying	Oven
6. Permeability of drying	Forced air heater
7. Mix pelletizing and pressing	High-intensity mixer or lab-scale spray dryer
	Laboratory press
8. TE and gradient furnace dilatometer	Gradient furnace
9. Properties of extruded ceramic	Hydraulic testing machine
10. Properties of pressed ceramic	Hydraulic testing machine
11. Mercury porosimetry and SEM	Mercury porosimeter
	Scanning electron microscope

of having safety rules meeting criteria of any and all local regulatory agencies and insurance requirements.

- All students must follow safety policies at the location of experiments, and the laboratory where experiments are conducted must be in compliance with all local and national safety rules.
- All students must wear approved safety glasses during laboratories. Students will not handle any chemical species during any experiments. The only exception is natural clay material and/or formed or fired clay objects.
- Students will wear dust masks when in the proximity of any mixers used in the labs and when handling fine particulate materials (clays).
- Students will be aware of mechanical hazards, including pinch points, during operation of extruders or presses and take precautions as expressed by local authorities.
- Students will use care when cutting extruded bars.
- Students will listen to instructions before operating any equipment.

9.3 Conducting Laboratory Exercises

The following suggestions are made for academic courses:

- Students may work in teams of four or five.
- Laboratory exercises should be completed and turned in after each lab is completed (some take a week to complete).
- A master data table is useful and should be made available to all students.
- Midterm and final exams concentrating on data manipulation and interpretation force students to ask questions about what information means and why certain trends are observed.

EXPERIMENT 1: PARTICLE SIZE (SCREEN ANALYSIS)

9.4 Purpose

To determine the particle size of raw materials used in brick (or other ceramic) manufacturing with particular emphasis in this experiment on determining the noncolloidal or coarser size fractions. You will first determine the moisture content of the as-received (nondried) material.

Then you will obtain a dried specimen of the same material (provided to you) and you will run a dry screen analysis, which will reflect the state of agglomeration of the material. The dried material will be poured on a screen deck with screens progressing from larger openings (top of deck) to smaller openings (bottom of deck). You will determine the net weight of the material caught on each screen.

Since as-received raw materials are agglomerated, it is necessary to break down the agglomerates in order to find the true particle sizes. To do this, you will slurry or *blunge* the material using a high-intensity mixer. You will pass this slurry over a screen to separate the particles into +200 mesh and –200 mesh fractions.

The +200 mesh fraction caught on the 200 mesh screen will be used next week in lab. In a like manner, the –200 mesh fraction caught in the bucket will be used next week in lab. We will use a dry sieve analysis of the cake on the +200 mesh fraction next week to get the true (unagglomerated) size distribution. We will use the –200 mesh fraction to obtain the subsieve particle size distribution using a sedimentation technique.

9.5 Procedure

To find the moisture content of as-received material (brick clay), weigh a small sample of your raw material (~30 g) and place on a piece of preweighed aluminum foil. Use masking tape to mark your name on your specimen. Place the pan in the oven for 45 min to remove moisture. Weigh the foil and clay again. Compute the moisture content as follows:

Tare weight of aluminum foil = _______ g.

$$\text{Raw material moisture (\% wet basis)} = \frac{\text{Loss in weight on drying (g)}}{\text{Initial weight before drying (g)}} \times 100$$

where loss in weight = (original weight – tare weight) – (dry weight – tare weight).

Moisture content, % WB: _______

Compute the moisture content on a dry weight basis:

$$\text{Raw material moisture (\% dry basis)} = \frac{\text{Loss in weight on drying (g)}}{\text{Dry weight of materials (g)}} \times 100$$

Show your calculation and answer below:

Moisture content, % DB: _______

9.6 Screen Analysis of Dried (Naturally Agglomerated) Materials

Weigh out 50 g of the dried raw material. Charge it to the screen deck using the screens in the chart. Run the deck on the sieve shaker for a minimum of 10 min. CPFT stands for cumulative percent finer than, and

$$\text{\% retained} = \frac{\text{Weight retained on a particular sieve (screen) size}}{\text{Total weight charged to screen deck } (\approx 50\text{ g})} \times 100$$

Screen	Opening (mm)	Weight retained (g)	% retained	CPFT
12				
20				
30				
50				
100				
Pan				
Pan				

9.7 Preparation of a Specimen for Screen Analysis after Washing (Not Agglomerated) and for a Sedimentation Test (Experiment 2)

Make a sample for the screen analysis by weighing out the quantity of raw material determined by the equation:

$$\text{Sample size (g)} = \frac{50\text{ g} \times 100}{100 - \text{moisture (\% DB)}} \times 100$$

Sample size = _______ g

Slurry or blunge the material by placing the sample of your raw material in 300 mL of water in a milkshake cup. Add 1 g of tetrasodium pyrophosphate (tspp) to the cup. As an alternative, you can use 1 g of automatic dishwasher detergent when you do this experiment. Mix for 1 min. Check your slurry to determine if lumps exist. If so, mix for an additional 1 min.

Some alluvial clay raw materials — particularly those based on montmorillonite ("gumbo" clays) — may require sitting overnight in the milkshake cup followed by 1 min of remixing. If your clay doesn't break up in the mixer, you probably have a gumbo clay.

Pour the residue over a prewet 200 mesh screen (0.075 mm or 75 μm opening), catching the residue in a bucket. Use a little extra water so you are sure to wash out all of the colloids (very small particles).

Place the screen in the lab oven with an identifying label to dry overnight. Save the residue in the bucket for the sedimentation analysis. Put your name on the bucket on masking tape. Cover the bucket with aluminum foil.

For your information:

Screen (U.S. sieve series)	**Preferred screen**	**Opening (mm)**	**Opening (μm)**
4 or 6	4 (can be omitted)	4.75	4750
12 or 14	12	1.40	1400
14 or 20	20	0.850	850
28 or 30	30	0.600	600
48 or 50	50	0.300	300
65 or 100	100	0.150	150
200	200	0.75	75

9.8 Questions

1. What is the difference in the moisture content on a dry basis as compared to a wet basis?
2. Why do materials such as brick clays exhibit a high degree of agglomeration when mined?
3. What is the function of tspp in making the slurry?
4. What is the expected difference in dry screen analysis when comparing as-received dried material to slurried and dried materials.
5. Does the technique used in this lab guarantee that the slurried dried screen analysis will reveal the ultimate (unagglomerated) particle size distribution?

EXPERIMENT 2: PARTICLE SIZE (SEDIMENTATION ANALYSIS TO DETERMINE COLLOIDAL SIZES)

9.9 Purpose

The purpose of this experiment is to determine the sizes of particles in the subsieve or colloidal size ranges. Many engineers classify soils as:

- Sand: Particle sizes greater than 200 mesh (75 μm).
- Silt: Particle sizes less than 75 μm and greater than 2 μm.
- Clay: Particle sizes less than 2 μm.

Since the finest sieve or screen size commonly employed is 325 mesh (opening 44 μm or 0.044 mm), particles smaller than 44 μm are usually called subsieve particles.

Subsieve particles are very important in ceramic production. They contribute to the plasticity and thereby the forming of the material. They have the most profound effect on drying and firing shrinkage of all particle sizes. Many companies perform daily sedimentation tests to ensure that they do not have wide variations leading to high scrap rates. Some companies consider sedimentation test results in evaluating new raw material sources. Sedimentation tests are also important since the colloidal material affects brick color.

9.10 Procedure

Obtain the cake or +200 mesh material from wet screening (Experiment 1). Break up the cake with a mortar and pestle. Preform a dry screen analysis as in Experiment 1.

Weigh the dried sample of +200 mesh material: ______ g.

Fill in the information below:

Screen opening	Weight retained (g)	% retained	CPFT
12			
20			
30			
50			
100			
Pan			

9.11 Start the Sedimentation Analysis

Stir the material in the bucket vigorously to insure that a cake is not left on the bottom of the bucket when you transfer material into the graduated cylinder. Now quickly transfer the -200 mesh slurry from the bucket into a 1000 mL graduated cylinder. Fill the cylinder to the 1000 mL mark with extra water. Use your hand as a stopper on the open end of the cylinder and invert the cylinder several times to mix the material. Be careful! If you spill the contents, you have to start over and you have to use the mop to clean up your own mess!

Use a soils hydrometer (No. 152H) to monitor the specific gravity of the unsettled portion of the slurry in the graduated cylinder. Take readings at the immediate beginning of the experiment and at the indicated times after you start.

Starting time: ______

Initial specific gravity: ______ g/L

Weight of solids: ______ g/L × 1.0 L = ______ g

Elapsed time (min)	Time planned for measurement	Actual time of measurement
2		
5		
15		
30		
60		
150		
1440		

Note: Between hydrometer readings, the hydrometer should be removed, rinsed, and placed back in the package or in a cup of deionized water. Don't leave the hydrometer continuously in the graduated cylinder. The hydrometer is calibrated for 68°F.

The hydrometer gives the number of grams per liter still in suspension. For example, a hydrometer reading of 46 g/L in ~1000 mL means that 46 g of material is left in suspension.

If you started with 100 g of particulate matter in the slurry, then at the time you take a reading of 46, the total amount settled out at that time is 100 – 46, or 54 g.

The Stokes Law equation to complete the data table is given at the end of this experiment.

9.12 Complete the Table

Starting time: ______

Initial hydrometer reading: ______

Actual elapsed time (min)	Hydrometer reading (g/L)	Particle diameter (μm)	CPFT (Note 1)	% passing for the total sample (Note 2)

Note 1: Calculate CPFT for the –200 mesh particles by:

$$\frac{\text{Hygrometer reading at time } t}{\text{Initial hygrometer reading}} \times 100$$

Note 2: For the total washed sample, simply multiply each entry for a particular size in the column % passing by the ratio:

$$\frac{50 \text{ g} - (\text{Weight retained on 200 mesh sieve})}{50 \text{ g}}$$

to get the corresponding value in the last column, that is, the percentage passing for the total sample in both experiments. This is the same as using the hydrometer reading in g/L × 1 L to give grams in suspension. Simply dividing by the starting weight of raw material (× 100) gives the CPFT.

9.13 Using Stokes Law to Calculate the Size of a Particle Falling at Any Time

It is easy to use a calculator to find the size of particle falling at any time during sedimentation using the formula (known as Stokes Law):

$$V = \frac{2\left(\rho_s - \rho_w\right)gr^2}{9\eta} \tag{9.1a}$$

where: r = the radius of particle falling at a given time in centimeters (centimeters × 0.0001 = micrometers).

η = the viscosity of water in poises (0.010 poise or 0.010 g/s-cm at 20°C).

V = the velocity of the particles in cm/s (use the half height of the water column/time).

ρ_s = the density of clay (2.65 g/cm^3).

ρ_w = the density of water (0.99823 g/cm^3 at 20°C).

g = the acceleration due to gravity (981 cm^2/s).

With the values plugged in, the equation simplifies to:

$$V = \frac{2\left(2.65 - 0.99823\right)\left(981\right)r^2}{9\left(0.01\right)} \tag{9.1b}$$

or

$$V = 36008r^2 \tag{9.1c}$$

Recall that velocity is just distance divided by time, that is:

$$V = \frac{x}{t} \tag{9.2a}$$

Since the object of a sedimentation experiment is to determine the weight left in suspension (converted to CPFT) as a function of particle diameter *(d)* falling at any particular time, the Stokes Law equation can be rearranged as follows where x is the half-height of the sedimentation container and t is sedimentation time in seconds:

$$d = 0.1354\left[\frac{\eta}{\rho_s - \rho_w}\left(\frac{x}{t}\right)\right]^{1/2} \tag{9.2b}$$

Or, utilizing the values for clay and water for room temperature (20°C) and recognizing that the average path of a particle is the half-height of the sedimentation cylinder of 16.5 cm:

$$d = 0.1354\left[\frac{0.010\ \text{g/s-cm}}{2.65 - 0.99823}\left(\frac{16.5\ \text{cm}}{t\ (\text{s})}\right)\right]^{1/2} \tag{9.3}$$

Stokes Law for suspension of clay in water at 20°C becomes:

$$d = 0.1354\left(\frac{0.0428}{t^{1/2}}\right) \tag{9.4}$$

where: d = the particle diameter in centimeters (cm).
t = time for sedimentation in seconds.
x = the half-height of the sedimentation chamber (16.5 cm).
If time is 120 s, then the solution is:

$$d = 0.1354\left(\frac{0.0428}{120^{1/2}}\right) = \frac{0.0428}{10.9544} = 0.0039\ \text{cm} = 0.039\ \text{mm} = 39\ \mu\text{m}$$

Example: Slurry 50 g of dry clay in water forming 1 L (1000 mL) of suspension. At 2 min (120 s), the hydrometer reading is 45 g/L. The cylinder is full to a height of 30 cm.

Therefore, the average particle is falling 15 cm. The average velocity is 15 cm/120 s or 0.125 cm/s. Plugging these values into the formula, we get:

$$d = \left(\frac{15}{120}\right)^{1/2} = 0.0037\ \text{cm} = 37\ \mu\text{m}$$

So, from the specific gravity data we know that 45 g/50 g × 100 or 90% of the raw material is finer than or equal to 37 μm, or CPFT = 90% at 37 μm.

Equation 9.2a can be corrected for other temperatures using the following data:

Temperature	Water viscosity (P = g/s-cm)	Water density (g/cm³)
10°C (58°F)	0.013	0.99973
20°C (68°F)	0.010	0.99823
30°C (86°F)	0.008	0.99567

9.14 Using Sedimentation Analysis Data

Many brick companies will follow this procedure and check only the percentage of 2 μm particles in their clay shed. The percentage of 2 μm material can be conveniently found by obtaining only the starting slurry specific gravity and the 5 h hydrometer reading. Then the percentage of −2 μm material can be charted with dryer loss and firing loss. In the event that excessive −2 μm material is found, clays can be blended or selective mining can be employed.

The data can be plotted using techniques in the Experiment 1. The distribution modulus *(n)* can be used as a single number to predict extrusion and drying behavior.

9.15 Producing a Total Particle Size Distribution

You can produce a total particle size distribution from the data in this experiment and Experiment 1 by filling in the data table on the next page.

Sieve	Sieve opening (µm)	% retained	CPFT

+ Sedimentation data

Particle diameter (µm)	CPFT for total sample

9.16 Example Problem for Total Size Distribution

The brick raw material is from Colorado; 50 g dry weight and 22 g dry material are retained on the 200 mesh sieve after washing.

Screen analysis:

Sieve	Opening (μm)	Weight retained (g)	% retained	CPFT
4	4750	0	0	100
12	1400	0	0	100
20	850	5.8	11.6	88.4
30	600	2.8	5.6	82.8
50	300	3.7	7.4	75.4
100	150	5.1	10.2	65.2
200	75	4.6 (Total weight of 22.0 g on all screens)	9.2	56.0

Sedimentation test (initial hydrometer reading: 28 g/L):

Time (min)	Hydrometer reading (g/L)	Diameter (µm)	CPFT (this sample)	% passing (total sample)
2	26	37	92.9 (26/28 × 100)	52.0 (Note 3)
5	25	24	89.3	50.0
16	23.5	13	83.9	47.0
30	23.0	10	82.1	46.0
61	21.5	7	76.8	43.0
212	17.5	3.8	62.5	35.0
1430	15.0	1.5	53.6	30.0

Note 3: Calculated as in Note 2: (50 – 22) / 50 × 92.9 = 52.0 or (26 g/L × 1 L) / 50 g = 52.0.

The data shows a fairly typical distribution for a ceramic material. Shale base mixes typically contain less colloidal material (especially with harder shales). Since the Colorado material is alluvial clay, it contains a higher fraction of colloids (–2 µm).

The distribution follows reasonably a distribution of the type:

$$CPFT = \left(\frac{x}{x_{\max}} \right)^n$$

where x is the particle size, x_{max} is the predicted maximum size in the distribution (sometimes called the fineness modulus or f_m), and n is the distribution modulus. In this case $n = 0.155$ and $\log x_{max} = 4$ or $x_{max} \sim 10{,}000$ µm (~0.375 in.).

The value of the distribution modulus of my material: _______

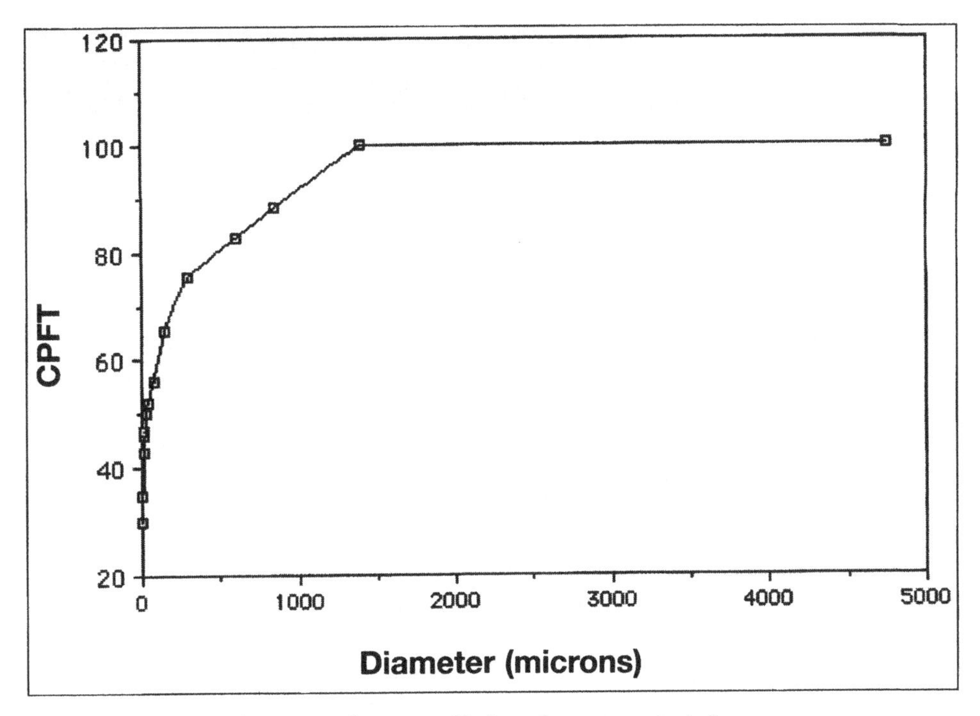

Figure 9.1: CPFT versus diameter (Colorado raw material).

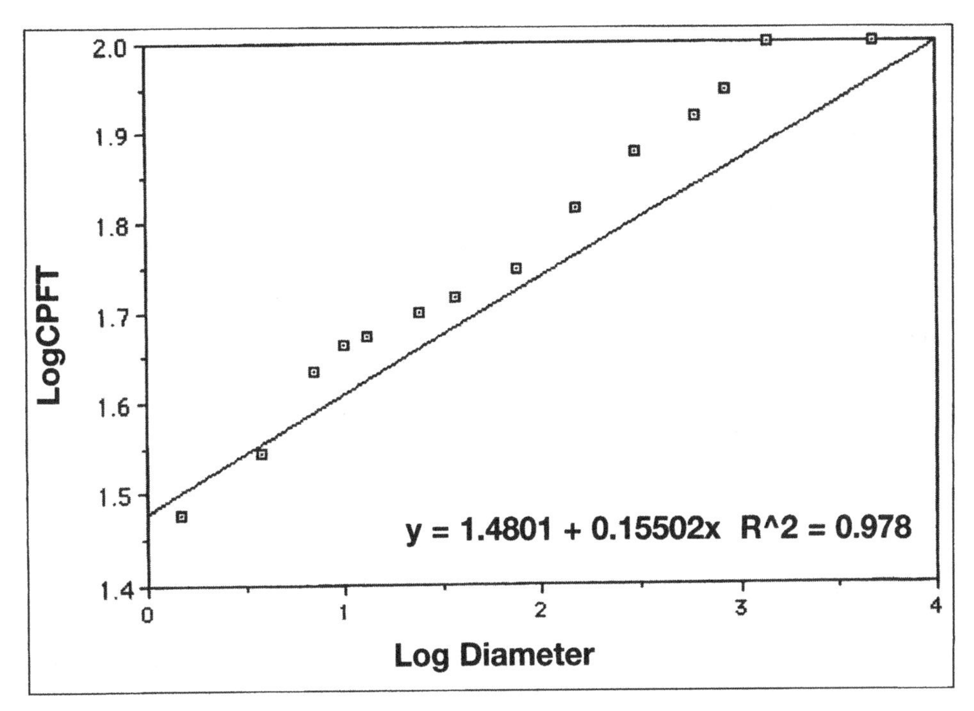

Figure 9.2: Log CPFT versus log diameter (Colorado raw material).

EXPERIMENT 3: MOISTURE ADSORPTION AND IGNITION LOSS

9.17 Purpose

The purpose of this experiment is to demonstrate basic properties that affect thermal processing of ceramics.

The first property is moisture adsorption (MA) of dried clay raw materials. The moisture adsorption is expressed as the percentage of weight gain on a previously dried specimen while held in a controlled-humidity atmosphere. The amount of moisture adsorption depends on the mix of minerals in the raw material and on the fineness of the mix. The value of moisture adsorption also indicates the tendency of a material to readsorb moisture if the product is dried and held in an ambient atmosphere prior to firing. High values of moisture adsorption can lead to dryer loss unless ceramic parts are immediately inserted into the kiln after drying. This is because moisture adsorption leads to expansion opening cracks and because steam spallation could occur if moist brick are inserted directly into the kiln.

The second property is the loss on ignition (LOI) of the raw material. This property is commonly reported as the percentage weight loss when a specimen of clay is heated to 1000°C (1832°F) and held for 1 h. It can be viewed as the weight change percentage when the product is fired. The weight losses in firing are due primarily to (a) loss of water chemically combined in the clay mineral (also called *dehydroxylation*), (b) burn out of carbon in the raw materials, and (c) decomposition of accessory minerals such as pyrite (FeS_2).

The ignition loss is an inexpensive and rough indication of the clay mineral type. Exact identification of the mineral types is by X-ray diffraction analysis.

9.18 Procedure

Each student will perform one MA and one LOI experiment. You will be assigned either a shale material used in producing red bricks or a bentonite clay.

Your assigned raw material is: ______________________________

9.19 LOI Experiment

Weigh a porcelain crucible provided by the lab assistants using only the top loading balance indicated for today's lab:

Weight of your porcelain crucible _______ g (Value 1)

Fill the porcelain crucible about half-full with your dried raw material (Do not pack it down). Identify your crucible using a marker pen. Weigh the crucible containing the material.

Weight of crucible + dry material: _______ g (Value 2)

Conduct the experiment and record the following:

Weight of crucible and ignited material: _______ g (Value 4)

9.20 MA Experiment

Obtain a polyethylene cup or the designated container and weigh it to the nearest 0.01 g:

Weight of MA empty container: _______ g (Value 3A)

Dry weight of MA specimen plus container: _______ g (Value 3B)

Label your cup with a marker pen.

Place the MA sample in one of the coolers (a convenient enclosure) containing water and an excess of table salt (to create a constant humidity environment). The MA sample must be on a pedestal above the water line. If a brick is used to support the MA specimen, the top of the brick should be covered with aluminum foil to prevent moisture transmission through the brick pedestal into the MA specimen. Make sure that the cooler is closed after you put your specimen inside to preserve the controlled humidity environment. The lab assistants will weigh your cup.

9.21 Calculations

LOI:

Weight of LOI crucible after test: _______ g (Value 4)

Compute the LOI of the specimen:

$$\text{LOI} = \frac{\text{Value 2} - \text{Value 4}}{\text{Value 2} - \text{Value 1}} \times 100$$

LOI = ______ % (Value 5)

Characteristic LOI values are kaolinite (~13%); illite (6%); and montmorillonite (5+%).

My mineral type may be: _______________ and/or _______________.

MA:

MA Specimen Weight after Humidity Exposure: ______ g (Value 6)

Compute the MA by the calculation:

$$MA = \frac{\text{Value 6} - \text{Value 3B}}{\text{Value 3B} - \text{Value 3A}} \times 100$$

MA = ______ % (Value 7)

Most brick raw materials will readsorb 0.2–0.8% moisture from a humid atmosphere after being dried. If moisture readsorption exceeds about 1 wt%, defects could be introduced into the product if product is allowed to dwell after drying prior to charging the kiln (especially in the summer if the humidity is high).

EXPERIMENT 4: EXTRUSION OF CLAY BODIES

9.22 Purpose

Extrude low-, medium-, and high-water-content clay mixes using the vacuum extruder. Observe operating parameters and column stiffness (penetrometer reading) as a function of water content.

9.23 Procedure

Each group will extrude about 10 kg of clay at the assigned water content. The procedure will be to weigh the dry clay and charge it to the mixer.

Take caution operating the high-intensity mixer. Note that the rotor must be turned on and spinning before the pan is turned on. Add water slowly, allowing your instructor to help you gauge the desired consistency. Once the proper consistency is reached, turn off the mixer, remove the mixer pan, and remove your material to a clean plastic pail. Clean the mixer pan. Record the water quantity added to your material.

Operate the extruder with the assistance of the laboratory staff. Record readings on the extruder and use the penetrometer as required. Cut bars to a length of 6 in.

Mark the specimens for your group with your group letter and a number indicating the order in which the bars were made (1–8).

Incribe each bar on the top with the sharp end of the vernier caliper with the caliper set at 5.00 in. Weigh each bar and record its weight to the nearest 0.01 g.

Wrap each bar individually in aluminum foil and use tape or markers identify it on the outside of the wrapper. Place the wrapped bars on the pedestal in the assigned cooler with water in the bottom of the cooler, and keep the cooler lid closed. The cooler is used to create a 100% relative humidity atmosphere to prevent drying of the bars.

These bars will be used in later experiments:

- Nos. 1 and 8: Stages of drying and calculation of moisture content
- Nos. 2 and 7: Permeability in drying
- No. 3: Thermal expansion, gradient furnace. Must be dried beforehand.
- Nos. 4–6: Firing experiment. Must be dried beforehand.

9.24 Data

Enter the data for all groups:

Measurement or parameter	Group A: High water content	Group B: Medium water content	Group C: Low water content
Weight of dry red clay mix (g)			
Water added to mixer (mL)			
Moisture content (% WB)			
Vacuum level (in. H_2O)			
Extruder current (amp)			
Penetrometer reading			
Individual bar weights (g)			
Bar 1			
Bar 2			
Bar 3			
Bar 4			
Bar 5			
Bar 6			
Bar 7			
Bar 8			

9.25 Questions

1. Make a sketch of the vacuum extruder labeling key components.
2. What is the function of the chopper mechanism at the end of the feed auger?
3. Why did moisture content affect the penetrometer reading?
4. What differences are created in the properties of the extrudate by using vacuum as compared to extruding without the aid of vacuum?
5. Define the following extrusion defects and give the usual cure:
    a. Auger twist lamination
    b. Feather edging
    c. Residual stress defect seen after drying
    d. Die skin delamination
6. Name three fundamental differences in capability between a muller mixer and a high-intensity mixer (such as the Lancaster mixer).
7. Discuss how viscosity-modifying agents such as methylcellulose or Methocel may be used as a binder when extruding nonplastic materials.
8. Discuss how dispersants such as liganosulfonate (Additive A) affect product quality.
9. Draw and label the key components of a piston extruder.
10. When extruding with a piston extruder, discuss how extruder force varies with time during extrusion of a dilatent material.
11. Why is double-pugging beneficial in brick production?

EXPERIMENT 5: STAGES OF DRYING

9.26 Purpose

The purpose of this experiment is to monitor drying rate and shrinkage during the two stages of drying. Stage I is defined as the shrinkage period, where the drying rate must be slow enough to prevent cracking of the ceramic ware. Stage II is defined as the period where the remaining water is removed, where the temperature is elevated to remove any water left in capillaries or on the surface of colloidal particles.

9.27 Procedure

Each group will work on the specimens produced in the lab on extrusion. Use the designated two specimens (Bars 1 and 8) from Experiment 4.

Write the identity of your extrusion group: ____________

Penetrometer (A/B/C/D): ______

Remove the aluminum foil wrapping on the two bars to be used in this experiment. Measure the length, breadth, depth, and weight of both specimens for your group. Record the data on the next page. Note that you will use the shrinkage marks impressed in the specimen for length (gauge) measurements.

Place the specimens on the designated table top in front of the fan. Slightly elevate the specimen with spacers to allow air to flow underneath the specimen.

Divide the responsibility for measurements among your group: 1.5 h of drying), 18 h, 23 h, 25.5 h. Measure the weight and gauge length between shrinkage marks. Record the information in the data table. Then place the specimen in the designated oven.

Designate one student in your group to remove the specimens for your group from the oven and make the final measurement of gauge length and weight. Note any cracks in your specimen.

9.28 Data

	First specimen	Second specimen
Starting data (before drying)		
Length (mm)		
Width (mm)		
Depth (mm)		
Starting weight (g)		
Ending data (after oven drying)		
Oven dried weight (g)		
Dried gauge length (mm)		

Area exposed:

2 length × width + 2 length × depth + 2 width × depth =

Specimen 1 area exposed: _______ cm^2

Specimen 1:

Time planned	Actual time of reading	Elapsed time (h)	Weight (g)	Gauge length (mm)	Incremental change in weight (g)
Start					
1.5 h					
18 h					
23 h					
25.5 h					
Other					

After oven drying (230°F for 24 h minimum):

Weight: _______ g

Gauge length: _______ mm

Incremental change in weight: _______ g

The final weight of the sample after oven drying will be called the *dry weight (D).*

Specimen 2:

Time planned	Actual time of reading	Elapsed time (h)	Weight (g)	Gauge length (mm)	Incremental change in weight (g)
Start					
1.5 h					
18 h					
23 h					
25.5 h					
Other					

After oven drying (230°F for 24 h minimum):

Weight: _______ g

Gauge length: _______ mm

Incremental change in weight: _______ g

9.29 Calculations

The total weight of moisture removed *(W)* in air and oven drying is:

W(g) = Green weight (g) – Dried weight (g)

The moisture content *(M)* of your product in the green state (%) is:

$$M \text{ (dry weight basis)} = \left(\frac{w \text{ (g)}}{\text{Dried weight (g)}}\right) \times 100$$

$$M \text{ (wet basis)} = \left(\frac{w \text{ (g)}}{\text{Green weight (g)}}\right) \times 100$$

These formulas can be used to express the moisture content at any point in drying by simply using the green weight at any one time. The weight of moisture at any point of interest *(W)* is the green weight minus the dry weight.

Moisture content of the green product:

Specimen 1: _______ % (mark wet or dry basis)

Specimen 2: _______ % (mark wet or dry basis)

Average of both: _______ %

The linear shrinkage *(S)* can be computed at any point in drying by using the formula:

$$S\ (\%) = \left(\frac{\text{Green gauge length} - \text{Gauge length at the time of interest}}{\text{Green gauge length}} \right) \times 100$$

Complete the tables:

Specimen 1:

Elapsed time (h)	**Linear shrinkage (%)**	**Moisture content (%) (Indicate wet or dry basis)**

Specimen 2:

Elapsed time (h)	Linear shrinkage (%)	Moisture content (%) (Indicate wet or dry basis)

If your sample achieved a constant length in air drying (i.e., no length change after placing it in the oven), you know the shrinkage and moisture content at the critical moisture content — the moisture content where shrinkage stops and you can pour on the heat. If you have this information, provide it below:

My critical moisture content is: ______ % (WB) or ______ % (DB)

At the critical moisture content, the linear drying shrinkage is: ______ %

Complete the table using the data you recorded for Specimen 1 during drying.

Elapsed time (h)	Weight (g)	Incremental change in weight (g)	Water evaporated per unit area (g of incremental weight/cm^2)

9.30 Graphing Results

Produce a graph of linear drying shrinkage versus moisture content for Specimens 1 and 2.

Produce a graph of drying rate versus drying time, fit an equation to that line, and provide your equation on the graph (using a computer utility such as Microsoft Excel™).

Produce a graph of average drying rate versus moisture content for each of the groups, labeling each curve with the identity of the group and the moisture content (DB) and penetrometer for that group.

EXPERIMENT 6: PERMEABILITY IN DRYING (MOISTURE MOVEMENT IN DRYING)

9.31 Purpose

The purpose of this experiment is to demonstrate a practical method for determining the rate of moisture movement through the body during drying and to show how different materials can exhibit different moisture movement rates. To learn the most from this experiment, it is necessary for students to compare their results on permeation rate (compare data from each penetrometer range for extruded products).

All shale brick or shale/kaolin brick typically exhibit a high moisture permeation rate reflecting the fact that they are easy to dry. Brick based on alluvial clay typically exhibit a much lower permeation rate, and may require additions like grog or sawdust to open up the body to facilitate drying (increase rate and reduce shrinkage).

Permeation rate depends on density and content of colloidal particles. In this experiment, our focus will be to determine if formed density (as reflected by penetrometer reading) affects permeation rate.

Instructor's Note: Extremely fine materials may show little difference in results, and grog additions or other pore-enhancing materials, such as wood flour, are recommended for some specimens to achieve differentiation in results in this experiment.

9.32 Procedure

Penetrometer reading during extrusion of my materials: _______

Cut Specimens 2 and 7 from your extrusion group as follows:

Cut off Section A before you begin the experiment. It should be cut about 1

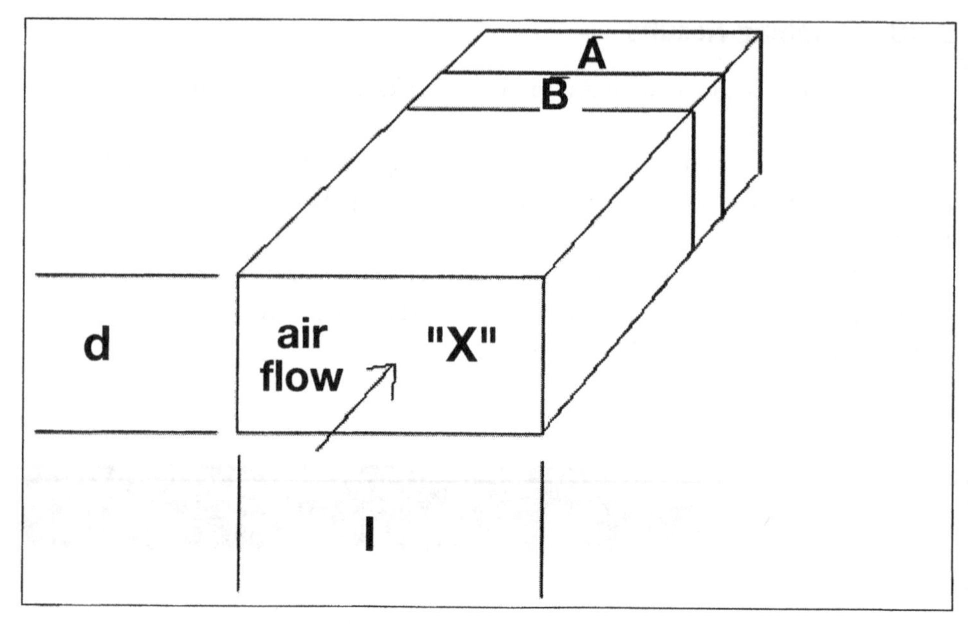

Figure 9.3: Specimen geometry in the permeation test.

in. from the end of the specimen. The purpose of Section A is to determine the starting moisture content of the specimen. *Do not cut off Section B until the end of the drying exposure!*

- Mark the specimen with your initials on Sections A and B as well as on the rest of the specimen. The specimen geometry will appear as in Figure 9.3.
- Weigh Section A and record the value on the data table (Value 1).
- Place Section A in the designated dryer for the specimen to dry at ≥230°F overnight.
- Weigh the remaining green specimens after removing Section A and record the value on the data sheet (Value 2).
- Measure length and diameter in millimeters and record the information on the data sheet (Value 2B).
- Wrap the specimen in one layer of aluminum foil so that only Face X is exposed. Use rubber bands to ensure a tight fit of the foil against the specimen.
- Weigh the wrapped specimens and record on the data sheet (Value 3).
- Place the wrapped specimens on the designated table in front of the heater.

To create our drying conditions, we will use a portable heater as a matter of convenience. Make sure that the airflow is directed toward Face X.

Data Table for Experiment 6

Name: ______________________________ Penetrometer: __________

Description	Value	Bar 2	Bar 7
Data:			
Green weight of Section A (g)	1		
Specimen green weight after removing A (g)	2		
l dimension of specimen after removing A (mm)	2B		
d dimension of specimen after removing A (mm)	2B		
Weight of wrapped specimen (g)	3		
Time started			
Time completed drying			
Total elapsed time in drying (h)	4		
Temperature near face *x* of specimen (°F)			
Weight of the wrapped bar at the end of drying (g)	5		
Green weight of Section B after partial drying (g)	6		
Dry weight of Section A (g)	7		
Dry weight of Section B (g)	8		
Calculated values:			
Moisture content of Section A (%DB)	10		
Moisture content of Section B (%DB)	12		
Permeation (% change) (%DB)	13 or 13A		
Area for evaporation (mm^2)	14		
Dry weight of remaining specimen (g)	15		
Permeation rate (g/m^2-h)	16		
Permeation rate (lb/ft^2-h)	17		

An alternative procedure is to use an extended air drying period directing ambient air toward the exposed drying face. A period of 3–5 days of air drying followed by an oven drying step has been successful with low-permeability (tight) bodies (compositions).

- Record the time started on the data sheet.
- Record the dry bulb temperature in the region of your specimen on the data sheet. This can be displayed on a thermocouple readout or on a thermometer at the drying station.
- The drying experiment should continue for at least 5 h for both of your specimens. A designated student from each group must be at the lab to complete the experiment.
- Record the time at which you end the experiment on the data sheet.
- Compute the total elapsed time in drying in hours and provide on the data sheet (Value 4).
- Weigh the wrapped bars at the end of the experiment and record the information on the data sheet (Value 5).
- Cut off Section B from the end of the partially dried specimens. Record their weights on the data sheet (Value 6).
- Place Section B in the designated oven. Be sure it is marked with your initials.

9.33 Second Day, Afternoon

Retrieve Sections A and B for both specimens from the oven. Obtain their weights and record on the data sheet: dry weight A (Value 7) and dry weight B (Value 8).

9.34 Calculations

Wet weight A (Value 1)	
– Dry weight A (Value 7)	
Difference (Value 9)	

Compute the moisture content of the green specimens using the data for Section A in each case for the two specimens.

Calculate for each specimen:

$$\text{Moisture content of A (dry basis)} = \left(\frac{\text{Difference Value 9}}{\text{Dry weight A Value 7}}\right) \times 100$$

Moisture content of A (dry basis) = ______ % (Value 10)

Wet weight B (Value 6)	
– Dry weight B (Value 8)	
Difference (Value 11)	

Compute the moisture content of the partially dried end of both specimens using the data for Section B.

Calculate for each specimen:

$$\text{Moisture content of B (dry basis)} = \left(\frac{\text{Difference Value 11}}{\text{Dry weight B Value 8}}\right) \times 100$$

Moisture content of B (dry basis) =______ % (Value 12)

Permeation is the quantity of water that was removed from the end of the specimen away from the airflow (representing the interior of a product during drying). This is simply the difference in the moisture content between Sections A and B, which is computed as follows:

Moisture content of A (dry basis): ______ % (Value 10)

Moisture content of B (dry basis): – ______ % (Value 12)

Permeation: = ______ % (Value 13)

Note: If your specimen crumbled in drying or you lost your specimen, permeation can also be calculated using Values 3 and 5 as follows:

Wrapped wet specimen (g) (Value 3)	
- Wrapped specimen after drying (g) (Value 5)	
Difference (Value 5A)	

Then the permeation is:

$$\text{Permeation} = \left\{ \frac{\text{Difference (Value 5A)}}{[\text{Wet weight specimen (Value 2)}][100 - \text{Moisture content of A (Value 10)}]} \right\} \times 100$$

Permeation = _______ (Value 13A)

The area for evaporation for each specimen is determined using Value 2B as follows:

Value l _______ mm × Value d _______ mm = _______ mm^2 (Value 14)

The dry weight of the remaining partially dried specimens (the permeation specimens) are calculated as follows:

$$\text{Dry weight perm. specimen} = \left[\frac{(\text{Wet weight of specimen})(100 - \%\ \text{moisture content of A})}{100} \right]$$

Dry weight permeability specimen = _______ g (Value 15)

This could have been obtained by drying the entire specimen, but we have chosen to perform the calculation in a more general way as might be appropriate for a larger product or a specimen with more complicated geometry.

The permeation rate (PR) is calculated as follows:

$$\text{PR} = \left[\frac{(\text{Permeation Value 13 or 13A})(\text{Dry weight perm. spec. or Value 15})(10{,}000)}{(\text{Area for evaporation or Value 14})(\text{Elapsed time or Value 4})} \right]$$

Permeation rate: _______ g of water/m^2/h (Value 16)

High values are typically around 2300 g/m^2-h; low values may be around 300–400 g/m^2-h.

To convert this to pounds per square foot per hour, multiply Value 16 by 0.000205:

Permeation rate: _______ lb of water/ft^2/h (Value 17)

9.35 Questions

1. If I have a low permeation rate, what can I do about it to speed up drying?
2. What effect does clay type have on permeation rate?
3. What effect does temperature have on water viscosity and permeation rate?
4. What effect does percent of colloidal particles (<2 μm diameter) have on permeation rate?

EXPERIMENT 7: MIX PELLETIZING

9.36 Purpose/Introduction

Most ceramic and materials engineers learn little about solid–liquid mixing techniques applicable to plastic or dry mixes in their university classes. The purpose of this laboratory is to demonstrate the superior dispersion capability of a high-intensity mixer (Eirich or Lancaster type) coupled with its ability to produce pellets.

Many old-fashioned mixers used in ceramic fields are muller types, which have a wheel that mashes the mix and a scraper to pull the material into the path of the wheel. These mixers could handle plastic materials, but they had very low shear and thus very low capability to disperse minute additions to the mix.

High-intensity mixers were developed by a German company, Eirich Machines, after 1950. They initially featured a high-speed impeller to break up the cake of material flattened by a wheel. These mixers came into widespread use in the United States and Japan in the 1960s. In the 1970s, inclined high-intensity mixers — which don't use a wheel — were invented. The inclined mixers made mix pelletizing particularly easy.

The theory of the intensive mixer for solids is that the high-speed impeller breaks up agglomerates so that small batch additions (typically down to 0.01 wt%) can be intimately mixed with or dispersed within a powder blend. This also allows complete dispersion of liquids, making old-fashioned aging of the batch obsolete.

Pelletizing is a process of agglomeration to make fine powders (typically <200 mesh) into a free-flowing powder by producing rounded agglomerates or pellets of diameter ~0.1–1.0 mm. Such pellets will readily flow into a press cavity or mold so that intricate parts can be made. Many familiar products, such as spark plugs, are made from agglomerated materials.

Pellets are made by two basic processes:

- Spray drying, where a slurry is sprayed into a cylindrical dryer, forming individual droplets that dry and form spherical pellets.
- Mix pelletizing, where a liquid is dispersed in a mix using the high-speed impeller, followed by a slow-speed rolling action to form the pellets.

The choice of mix pelletizing or spray drying may depend on the volume of pellets (prepared powder) necessary. Large-volume operations typically use spray drying.

Mix pelletizing is usually followed by a drying operation so that the pellets gain strength. Then the pellets may be sent over a scalping screen, with oversized pellets returned to the mixer.

One final advantage of the intensive mixers is their ability to dry back the batch if it is inadvertently overwet with water. By simply running the impeller (rotor) on high speed, the batch will rapidly dry out.

9.37 Procedure

Using the natural clay (red mix), charge the high-intensity mixer with a known weight of dry material (~2700 g).

Record the weight of dry material: ______ g

Add 18% water (dry weight basis) to the mixer.

Volume of water = ______ mL

Always turn on the rotor first to avoid unnecessary torque on the rotor when the pan is running. Use a low pan speed in this experiment.

Approximate angle of inclination of the mixer pan (from horizontal): ______°

Run the rotor on high speed for 30 s. Is the mix plastic? ______

Run the rotor on high speed for another 30 s. Is the mix plastic? ______

Run the rotor on high speed for another 30 s. Is the mix plastic? _______

Is the mix heating (getting hot)? _______

Continue high-speed agitation as dictated by your instructor.

Now switch the rotor to low speed. Mix for about 30 s until you see pellets. Scalp some of the the material into +16 mesh and –16 mesh fractions (1 scoop).

Weight of +16 mesh pellets: _______ g

Weight of –16 mesh pellets: _______ g

Compare dried –16 mesh pellets (made by the instructor) to spray-dried red mix.

Which pellets flow better? _______

Which are more round? _______

Turn the mixer on with the rotor on high speed.

How long does it take for the material to get hot? _______ min

EXPERIMENT 8: THERMAL GRADIENT FURANCE TESTING/ THERMAL EXPANSION

9.38 Purpose/Introduction

A thermal gradient furnace is a device that fires a sample for a defined dwell period under a known temperature gradient. This allows an experimenter to observe shrinkage and fired color as a function of temperature on the same specimen. This technique is particularly useful for products where fired color is an important customer acceptance criterion (brick, quarry tile, roofing tile, pottery, etc.).

Thermal expansion is a property measured on a device called a dilatometer (dilation means "change in length"). The unfired or fired ceramic specimen is inserted into the dilatometer, and a heating program is begun. The thermal expansion of the specimen may be measured in an air atmosphere, or a shielding gas (Ar, CO, CO_2, etc.) may be used during the test.

On heating, most ceramics will exhibit expansions and/or contractions during the process of heating. Some reasons are:

- Crystallographic conversions and inversions (for example, $\alpha \rightarrow \beta$-quartz at 573°C results in a 2% linear expansion on heating or a contraction of similar magnitude on cooling).
- Bloating above 950°C due to gas entrapment in a body containing glass where the body is not sufficiently permeable for the gas to escape.
- Shrinkage above the temperature at which sintering begins.
- Expansion on the formation of new minerals, such as $3Al_2O_3 \cdot 2SiO_2$, either by direct reaction (called *primary mullite*) or by recrystallization on cooling (called *secondary mullite*).

A typical thermal expansion plot for two clay minerals is shown in Fig. 9.4. One of the clays (lower curve) is an Oklahoma shale material exhibiting expansion in the range 550–600°C (due to the quartz inversion) with shrinkage beginning about 850°C. The other clay is a marl from Ontario (~15% limestone, 85% clay) that also exhibits an expansion in the zone where the quartz inversion is expected. In the case of the Canadian clay, shrinkage begins about 875°C as lime-rich glasses are formed. If heating is not too fast, the curve will form a small plateau at about 925°C as the following recrystallization reaction takes place:

$$\text{Calcium aluminosilicate glass} \rightarrow CaO \cdot Al_2O_3 \cdot 2SiO_2 \text{ (anorthite)}$$

In firing the product made from the Canadian clay, the heating schedule must be slowed in the realm of 875–925°C to prevent excessive shrinkage (and cracking). Once the anorthite is formed, the product is more volume stable (i.e., it will shrink less) and normal firing can continue. The phase equilibrium diagram for the system CaO-Al_2O_3-SiO_2 illustrates these processes.

9.39 Procedure

Retrieve your group's bar that was fired in the gradient furnace (Groups A, B, or C). Using the data for temperature versus position, fill in the chart below for data points on linear shrinkage versus (width) position.

Starting width: = ______ mm   Time in furnace = ______ h

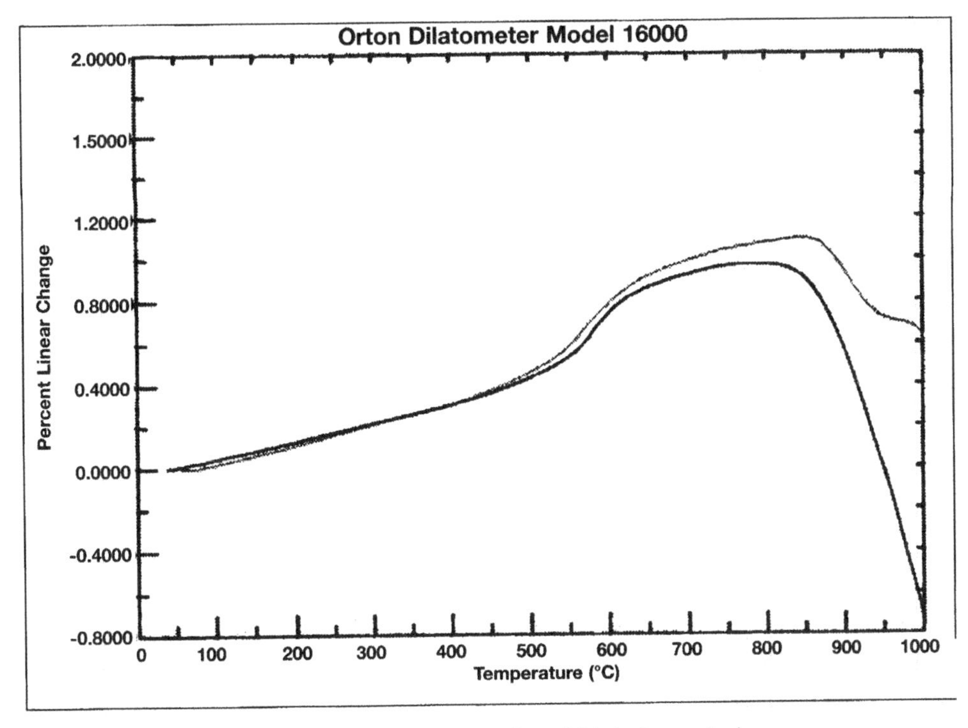

Figure 9.4: Thermal expansion of normal and high-lime shale.

Temperature	Width	Width shrinkage (%)

Make a graph of shrinkage versus temperature for your product using a computer to generate the graph. Attach your graph to your lab report.

In the lab we will run the thermal expansion on the dilatometer on the following products:

- Unfired bar specimen from Group C on heating and cooling
- Fired brick made from similar raw materials

Both of these will be run to about 1150°C.

9.40 Questions

1. When did you see the quartz inversion on heating (what temperature range in °C)?
2. When did the body start shrinking during heating?
3. When did you see the quartz inversion on cooling in the originally unfired specimen?
4. Does the fired body contain free quartz?
5. What was the approximate firing temperature of the fired product?
6. Calculate the thermal expansion coefficient in the following units for the fired product:

   ______ %/°C

   ______ /°C

   ______ /°F

7. Provide the thermal expansion coefficients (°C) of the following types of ceramics:

   Cordierite: ______

   Alumina porcelain (90% Al_2O_3): ______

   Mullite: ______

   Soda-lime-silica glass: ______

8. If you double the heating rate in a dilatometer test, what will be the effect on the observed temperature of the quartz inversion in a clay ceramic?

9. Why does glass typically exhibit a rapid increase in its dimensions prior to the transition temperature?
10. It is common for manufacturers to slow down firing schedules (on heating or cooling) whenever the product is in a zone where volume changes are expected within the product. Why?
11. A common defect in ceramics containing cristobalite (another form of crystalline silica) is *dunting*. Tell what dunting is.

EXPERIMENT 9: FIRING TRADITIONAL CERAMICS

9.41 Purpose/Introduction

In this lab our goal is to compare fired properties of bar specimens you made at various penetrometer readings and fired at two temperatures. You will learn about the following properties:

- Cold water absorption: the percentage increase in weight compared to dry weight after submersion in water at room temperature for 24 h. In many cases, this value is simply called *C*. Submersion for 24 h only partially saturates the ceramic (by filling capillaries).
- Boiled water absorption: the percentage increase in weight compared to dry weight after submersion in boiling water for 5 h. This is usually called *B*. This almost completely saturates the ceramic (fills most pores).
- Saturation coefficient: simply C/B. It is a number representative of the fraction of capillaries in the fired ceramic.
- Bulk density: the weight of the ceramic divided by its gross or bulk volume. The bulk density obtained by immersion techniques is sometimes called the *Archimedian density* because the Archimedes principle was used in its determination (weight of water displaced ~ bulk volume of specimen).
- Apparent bulk density: a value representing the density if there were no open pores.
- Apparent porosity: the percentage of volume of the fired ceramic occupied by pores.

Note that tests for these properties are described in ASTM Standards C-20, C-67, and C-133. We are running nonstandard tests because our specimen sizes are not standard, and our number of specimens tested is not the required number.

9.42 Formulas

$$\text{Saturation coefficient} = \frac{\text{Cold absorption \%}}{\text{Boiled absorption \%}}$$

For densities and apparent porosity, let:

d = dry weight.

w = wet weight (weight after boiling and then removing surface water with a wet cloth).

s = suspended weight.

Then:

$$\text{Bulk density} = \frac{w}{w - d}$$

in g/cm^3.

$$\text{Apparent density} = \frac{d}{d - s}$$

in g/cm^3.

$$\text{Apparent porosity} = \frac{w - d}{w - s}$$

in percent.

9.43 Data Table and Calculations

Fill in the data tables on the next two pages for all groups. Some data (in italics) were obtained for you to facilitate today's lab.

Save broken specimens for future labs.

Measurement	**Group A**		**Group B**		**Group C**	
Firing temperature (°F), if applicable	1930	2030	1930	2030	1930	2030
Gauge length, green (mm)						
Gauge length, fired (mm)						
Weight green (g)						
Weight fired (g)						
Fired length × width × depth (mm)						
Load in MOR test at failure (lb)						
Dry weight, 5 h boil (g)						
Wet weight after 5 h boil						
Dry weight, 24 h cold (g)						
Saturated weight, 24 h cold						
Suspended weight (g)						
Color observation						

	A		B		C	
Property	**1930°F**	**2030°F**	**1930°F**	**2030°F**	**1930°F**	**2030°F**
Moisture content of bars (% DB)						
Drying shrinkage of bars (%)						
Weight loss on firing (LOI) (%)						
Firing shrinkage (%)						
Cold absorption (C-67)* (%)						
Boiled absorption (C-67)* (%)						
C/B (C-67)* (%)						
Bulk density (C-20)* (g/cm^3)						
Apparent Density (C-20)* (g/cm^3)						
Apparent porosity (C-20)* (%)						
Modulus of rupture (C-67)* (report in lb/in^2 and MPa)						
Shade comments (light/medium/dark)						

9.44 Questions

1. What was the effect of firing temperature on fired density?
2. What was the effect of initial penetrometer reading on fired density and porosity?
3. Why is water absorption a preferred measurement in some industries instead of porosity measurement?
4. How did firing temperature affect strength (MOR)?
5. How would you determine how many MOR tests to run on one type or class of specimens to obtain 95% confidence in the data?
6. What was the MOR (in megapascals) of the Group A bars after firing to 2030°F?
7. What is the importance of measuring the *C/B* ratio of fired brick? (Hint: consider freeze-thaw durability)
8. Which has higher water absorption: a hand-made brick or an extruded brick, and why?
9. How many significant digits are there in a water absorption measurement?
10. How many digits to the right of the decimal point for an apparent porosity measurement are meaningful to a ceramic and materials engineer?

EXPERIMENT 10: FIRING DRY-PRESSED CERAMICS

9.45 Purpose/Introduction

In this experiment, you will determine the properties of two clay ceramics produced by dry pressing of commercially prepared and spray-dried clay. We will compare these properties to results from similar materials in Experiment 9.

Spray drying is a process used to agglomerate fine powders. The purpose is to render the powder free-flowing so it can easily fill a die cavity in a press. Uniform fill is very important in achieving the required pressed density, dimensions, and uniformity (within the pressed part). If you are unfamiliar with spray drying, it is expected that you will review Chapter 20 in *Principles of Ceramics Processing*, second edition, by James S. Reed, Wiley-Interscience or information in a comparable reference.

Spray-dried powder used in preparing the specimens used in this study was obtained from commercial production at a tile plant (or alternate spray dried raw material). In this plant, red body floor tile is made by pressing and processing tile through a roller hearth kiln. All tiles are glazed.

9.46 Procedure

To prepare the specimens used in this lab, the following steps were done in advance:

1. The surface of dried pelletized material was wet slightly using a water spray bottle. The laboratory technique involved spreading the powder out in a pan and lightly spraying the surface to wet the powder. This small amount of added moisture greatly improves the green strength of the specimens that are pressed.
2. One 2.25 in. diameter disc was produced by dry pressing at each of two pressures (Pellet 1 at 8000 lb force and Pellet 2 at 20,000 lb force).
3. The discs were fired to 1050°C and held for 1 h.
4. The green and dry dimensions and weights are given in the table below, along with data from preliminary Archimedian experiments.

In this lab exercise, you will complete the calculation of densities and absorption and run diametral compression test on fired discs

The formulas for densities, absorption, and apparent porosity per ASTM C 20 are given below. Let:

d = dry weight.

w = wet weight (weight after boiling and then removing surface water with a wet cloth).

s = suspended weight.

Then:

$$\text{Bulk density} = \frac{w}{w - d}$$

in g/cm³.

$$\text{Apparent density} = \frac{d}{d - s}$$

in g/cm³.

$$\text{Apparent porosity} = \frac{w - d}{w - s}$$

in percent.

$$\text{Absorption} = \frac{w-d}{d} \times 100$$

in percent. Note that the water absorption formula above produces a number comparable with the boiling water absorption *(B)* obtained in Experiment 9 (because w is after a 5 h boiling saturation).

For diametral compressive strength:

$$\sigma = \frac{2P}{Dt\pi}$$

where: P = applied load (on edge of disc) at fracture.
D = specimen diameter.
t = specimen thickness.

Prepared lab data (furnished by your lab instructor):

Description	Disc 1	Disc 2
Force in pressing (lb)		
Green diameter (mm)		
Green height (mm)		
Green weight (g)		
Dry diameter (mm)		
Dry height (mm)		
Dry weight (mm)		
Fired diameter (mm)		
Fired height (mm)		
Fired weight (g)		
Dry weight for density (g)		

Data to obtain in this lab exercise (use safety glasses when working in the vicinity of the mechanical testing machine):

Description	Disc 1	Disc 2
Wet weight (g)		
Suspended weight (g)		
Breaking load* (P)		

* in diametral compression test.

Save both halves of the diametral compression test specimens for future use.

9.47 Calculated Values

Property	Disc 1	Disc 2
Green density (g/cm^3)		
Dry density (g/cm^3)		
Drying shrinkage (on diameter) (%)		
Fired density (g/cm^3)		
Firing shrinkage (on diameter) (%)		
Apparent porosity (%)		
Absorption (%)		
Apparent density (g/cm^3)		
Diametral compressive strength (N/mm^2, MPa)		
Diametral compressive strength (lb/in^2)		

9.48 Questions

1. Explain why the density of the pressed specimen went up as pressure in forming increased.
2. Explain why the porosity of the pressed specimen went down when the pressure increased.
3. Compare green density values for extruded, fired specimens from Experiment 9 to pressed, fired specimens from this experiment.
4. Compare fired strength of the pressed and extruded specimens from Experiment 9 to pressed, fired specimens from this experiment. What factors limit your ability to compare these specimens? Name two.
5. Explain the basic process involved in spray drying.
6. What are the control variables in spray drying? Namc four.
7. What material properties influence ability to spray dry powders?
8. How is spray-dried powder charged to a mold box in a production press?

EXPERIMENT 11: ADVANCED PROPERTIES OF PRESSED AND EXTRUDED CERAMICS

9.49 Introduction

The purpose of this lab is to examine properties of ceramics made in prior labs by extrusion and dry pressing of similar materials. In this lab, we will use two techniques to look at specimens made by dry pressing (Experiment 10) of spray-dried red body tile mix and specimens made by extrusion of crushed and screened red body brick mix (highest penetrometer reading and highest firing temperature). We are particularly interested in techniques that reveal the *texture* of the ceramic parts. Texture is the spatial relationship, size, and quantity of microstructural artifacts including aggregate particles, pores, and relics from the forming processes (planar defects and residual agglomerates). We will use three techniques:

1. Scanning electron microscopy (SEM) with energy dispersive X-ray analysis (EDAX) capability. Surfaces will be viewed in secondary electron mode; that is, the image will be formed by detection of electrons ejected from the surface region of the specimen by the electron beam penetrating the specimen. Another mode of operation, detection of backscattered electrons from the beam itself, was not used in this analysis. EDAX involves detection of X-rays emanating from the specimen caused by electronic transitions within the surface area atoms on penetration of the electron beam.

2. Mercury intrusion porosimetry, which forces liquid mercury into the specimen, measuring the intruded volume versus applied pressure. By means of calculations, the pressure and intrusion data are converted to a pore size distribution.
3. Optical microscopy.

9.50 Suggested Data to be Acquired

- Photographs by SEM of surfaces of interest:

- EDAX printouts and graphs: mix pelletized powder (same raw material used in extrusion) and spray-dried powder (same raw material used in pressing).

	Column 1	Column 2	Column 3	Column 4
Row 1	Mix pelletized powder, 27×	Spray dried powder, 130×	Mix pelletized powder, 25×	Mix pelletized powder, 50×
Row 2	Extruded with lamination, 22×	Extruded, 1500×	Extruded, 350×	Pressed, 1500×
Row 3	Pressed 750×	Pressed 250×	Pressed 50×	Pressed, 90×

- Incremental intrusion graphs: extruded product with median pore size of ~1 μm and dry-pressed product with median pore size of ~20 μm.
- Summary sheets from mercury porosimetry experiments (notice bulk density, apparent density, and apparent porosity data).

9.51 Questions

1. Aside from the obvious difference in the size of mix pelletized and spray-dried pellets, what are the implications of shape of agglomerate on flow of the powders during pressing?
2. What are the implications of uniformity of texture on fracture when looking at the extruded specimen fracture surface (at 22×) and the pressed specimen fracture surface (at 50×)? What are the implications of uniformity of texture on strength?
3. From the SEM photographs, can you infer something about pore shape in extruded versus pressed specimens?

4. In the 90× photograph of the pressed (tile) specimen, some people can see a relic of the spray-dried powder in the fracture surface. Explain why you can see this.
5. Comparing the EDAX printouts of the fracture surfaces, you see more potassium in the tile specimen. How would this be of interest in fast firing of tile in a roller hearth kiln?
6. What is the fundamental reason that the pores are larger, on average, in the pressed specimen than in the extruded specimen?
7. There are apparently some pores in the extruded specimen in the area of 100–200 μm in size. Why would these be in the extruded specimen and not in the spray-dried specimen?
8. Why would a density calculated from the mercury intrusion data be more accurate than a density obtained using Archimedian data (suspended weight after 5 h boil in water, etc.)?
9. In firing of the specimens to a higher temperature, what would happen to the average pore size?
10. In firing either of the specimens to a higher temperature or longer time, what would happen to the value of the apparent porosity?

APPENDIX 1

Symbols, Terms, and Units of Measurement

Term	Quantity/definition	English dimension	Metric dimension
$\propto$	"Is proportional to"		
γ	Surface tension	lb/in.	dyne/cm
η	Viscosity	lb/ft-h	poise (dyne-s/cm^2)
ρ	Density	lb/ft^3	g/cm^3
A	Area	ft^2	m^2
Btu	British thermal unit	See definition	See definition
C	Temperature in degrees Celsius		
C_p	Heat capacity	Btu/lb-°F	cal/g-°C
DB	Dry weight basis	%	%
d	Particle diameter		
dT or ΔT	Change in temperature	°F	°C
F	Temperature in degrees Fahrenheit		
f	Applied or imposed force		
G	Gibbs free energy	Btu/lb mol	kcal/mol
H	Absolute humidity	lb of water per lb. of dry air (or grains of water per lb of dry air)	g of water per kg of dry air
H_s	Absolute humidity at saturation	lb of water per lb of dry air	g of water per kg of of dry air
H_p	Percentage of humidity	%	%
H_r	Relative humidity	%	%
h	Enthalpy	Btu/lb	cal/g or kJ/kg

Term	Quantity	English dimension	Metric dimension
Δh_v	Enthalpy of vaporization		
K	Temperature in Kelvin (absolute scale)		K
h_v	Enthalpy of vaporization	Btu/lb	kJ/kg
h_l	Latent enthalpy	Btu/lb	kJ/kg
h_s	Sensible enthalpy	Btu/lb	kJ/kg
h_{sat}	Enthalpy at saturation	Btu/lb	kJ/kg
h	Time (hours)	h	h
h	Heat transfer film coefficient	Btu/h-ft^2-°F	kcal/h-m^2-°C
m	Distance (meter) quivalent to 3.28 ft		
m	Weight (in physics, m denotes mass where $m = w/g$; w = weight and g = gravitational constant)	lb (pound)	g (gram)
M	Molecular weight		
M_c	Critical moisture content	%	%
μm	Distance (micrometer) equivalent to 3.66 microinch or 10^{-6} m		
MA	Moisture adsorption	%	%
n	Distribution modulus in the Dinger-Funk or Andreasen equations describing particle size distributions		
p	Pressure	lb/in.^{2}	N/mm^2
p_w	Partial pressure of water vapor in a humid air mixture	lb/in^2	N/mm2
p_a	Partial pressure of air vapor in a humid air mixture	lb/in^2	N/mm^2
p_t	Total pressure in a gas mixture	lb/in.^{2}	N/mm^2
q	Heat flow	Btu/h	cal/h
R	Universal gas constant	Btu/°F-lb mol	cal/K-mol
S	Entropy	Entropy units	Entropy units
V	Volume	ft^3	m^3
T	Temperature (also dry bulb temperature)	°F	°C
ΔT	Temperature difference	°F	°C
T_d	Dew point temperature	°F	°C
T_w	Wet bulb temperature	°F	°C
W_A	Weight of dry air in a humid air mixture	lb	kg
W_W	Weight of water vapor in a humid air mixture	lb or grains	kg
WB	Wet weight basis	%	%

APPENDIX 2

Conversion Factors

Multiply	By (Or operation noted)	To Obtain
BTU	251.63	cal.
BTU/lb.	2.326	J/g
BTU/lb.	0.555	cal/g
BTU/lb.	2.929×10^{-4}	KWh/lb.
BTU/lb.-°F	1	Cal/g-°C
°C	1.8 and add 32°	°F
cal.	0.003974	BTU
Cal/g-°C	1	BTU/lb.-°F
°F	Subtract 32° and multiply by $^5/_9$	
J/g	0.239	cal/g
J/g	0.430	BTU/lb
J/g	2.778×10^{-7}	KWh/g

APPENDIX 3

Answers to Selected Problems

CHAPTER 1

Problem 2

(a) Moisture DB = 6.38%. Moisture WB = 6.00%.

(b) The weights of the dish were different in the two determinations, suggesting a weighing error on the initial or on the final weighing determination.

Problem 5

(a) Plasticity begins as viscosity increases above 2% water addition with maximum plasticity at ~3.5% water addition. Above 3.5% water addition, viscosity drops, indicating particle separation. On drying, the material likely would stop shrinking in the interval 2–3.5% water content DB. A reasonable estimate is that shrinkage might stop at about 3% water content DB.

(b) The pore and capillary water is likely completely eliminated on drying to about 0.5–1.0% DB, that is, at the maximum moisture absorption value for traditional clay materials.

Problem 7

The ratio of the bulk to true density provides the fraction of solid matter in the porous ceramic (ρ/ρ_t).

The fractional porosity of the ceramic is ($1 - \rho/\rho_t$).

The percent of porosity in the ceramic is 100 $(1 - \rho/\rho_t)$.

The heat capacity of the porous ceramic then becomes:

$$C_p = \left(\frac{\rho}{\rho_t}\right) C_{p,\text{true}} + 100\left(\frac{1-\rho}{\rho_t}\right)(0.24 \text{ cal / g°C})$$

Problem 9

(a) The final temperature in the enclosure is 57.2°C (135°F).

(b) The sensible heat lost by the air in the container was 118,004 cal.

(c) The latent heat gained by the air in the container was 118,004 cal. (The heat exchange was adiabatic for all practical purposes.)

CHAPTER 2

Problem 2

The height of the kerosene is ~1.25 in.

Problem 8

(a) Mole fraction = 0.0243.

(b) Percent by weight = 1.52.

(c) PPM_w = 15494.

Problem 9

H_r = 50.9%.

CHAPTER 3

Problem 3

(Velocity at 60°C) / (Velocity at 20°C) = 2.14.

Problem 4

M_c occurs at 5 h into the drying cycle.

Problem 8

(a) Temperature and heat transfer conditions; that is, air circulation (during Stage II).

(b) Particle size distribution (affecting the path for diffusion) and green density (affecting the porosity thereby affecting diffusion).

CHAPTER 4

Problem 4

(b) 15.76 Btu/lb.

(d) 8.20 Btu/lb.

Problem 8

T = 152°F, H = 0.0188 lb/lb, H_r = 10.8%, V = 16.45 ft^3/lb, T_w = 90.6°F

Problem 10

48.61 lb/min.

CHAPTER 5

Solutions from calculation utilities may vary slightly from those below.

Problem 1

(a) 17 Btu/lb air.

(b) 0.

(c) 9.7 Btu/lb air.

(d) 0.0091 lb H_2O/lb air.

(e) 7.3 Btu/lb air.

(f) 9.7 Btu/lb air.

(g) 65°F.

(h) 0.

Problem 2

(a) 9.7 Btu/lb air.

(b) 7.3 Btu/lb air.

(c) 0.0066 lb H_2O/lb air.

(d) 59°F.

(e) 0.

Problem 3

(a) 7.1 Btu/lb air.

(b) 6.3 Btu/lb air.

(c) 0.0067 lb H_2O/lb air.

(d) 3.6 Btu/lb.

Problem 4

(a) 110°F.

(b) 117°F.

(c) 27 Btu/lb air.

Problem 5

(a) 89°F.

(b) 113°F.

(c) 89°F.

(d) 41 Btu/lb air.

(e) 52%.

(f) 105 lb air/min.

(g) 5%.

(h) 80%, 42%.

Problem 6

(a) 141°F.

(b) 76°F.

(c) 76°F.

(d) 76°F.

(e) 0.015 lb H_2O/lb air.

(f) Goes into kiln.

(g) Condensation.

(h) 47%.

Problem 7

(a) 32%, assuming production dropped from 0.022 to 0.015. However, Problem 6 removed only 0.015 lb of water even though 0.022 lb were fed into the dryer. Actually, there is no reduction in rate, but in Problem 7 the product was dried completely while in Problem 6 water was left in the product after the drying cycle (i.e., the product wasn't completely dried in Problem 7).

(b) 76°F.

(c) 76°F.

Problem 8

(a) 1.7 lb/min.

(b) 70%.

(c) 45.6.

(d) 41.

Problem 9

(a) 110°F.

(b) 110°F.

(c) 0.022 lb H_2O/lb air.

(d) Goes into kiln.

Problem 10

(a) 144°F.

(b) 110°F.

(c) 0.015 lb H_2O/lb air.

(d) 11.9 Btu/lb air.

(e) 164%.

(f) 634 (666 Btu/1000 brick).

(g) 163 (636 Btu/1000 brick).

(h) 471 ft^3 gas.

CHAPTER 6

Problem 1

(a) 1360 lb/min.

(b) 39,856 cfm.

(c) $T = 266°F$.

Problem 13

Long-term exposure to gases containing dilute sulfuric acid may lead to corrosion and failure of structural steel members.

Problem 17

Reset one kiln car after it has exited from the dryer such that the bottom set and top set brick as passed through the dryer are not switched in position. If the cracking pattern on this special car is the reverse of what has been seen exiting the kiln in normal production, the problem originates in the dryer.

CHAPTER 7

Problem 3

Drying in Stage I may be taking place from the ends of the specimen faster than from the circumference (suggesting that parts may be stacked horizontally on one another). The objects may be placed on end with an air gap between adjacent parts, thus allowing for uniform drying. The setter trays may be perforated in order to allow uniform drying of the set end.

Problem 5

The extrusion die is unbalanced, leading to residual stress in the cut bricks. This stress is relieved by deformation on drying of unrestrained bricks or cracking of restrained bricks.

Problem 7

The problem is auger twist lamination. Solutions include reduction of the extrusion rate, increasing the moisture content of the green batch, and adding a plasticizer to the mix composition.

CHAPTER 8

Problem 6

(a) Improve particle packing to achieve a distribution modulus $n = 0.2 - 0.4$ (Eq. 3.2).

(b) Add grog to the mix to reduce shrinkage.

Problem 7

The green product must be more permeable to steam or the drying cycle must become slower during this period. From a product standpoint, the particle size distribution can be adjusted to slightly increase porosity to minimize steam buildup.

Problem 9

To complete drying in a timely manner, add hot air during the last stage to the dryer chamber to achieve convective drying.

APPENDIX 4

Additional Psychrometric Charts

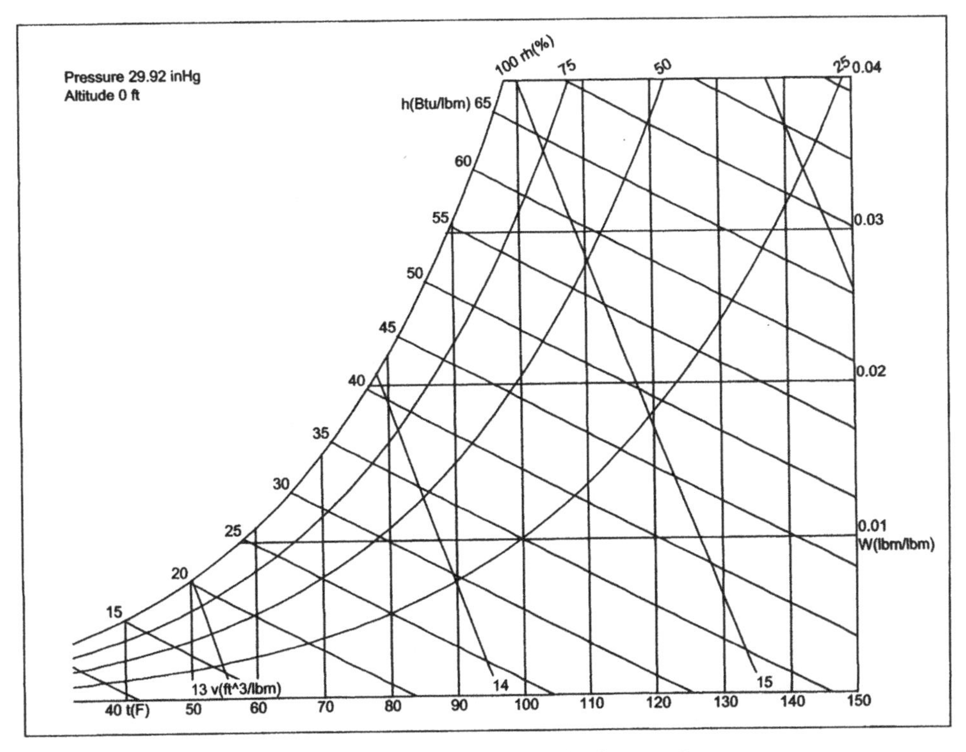

Figure A4.1: Low-temperature chart (Fahrenheit degrees).

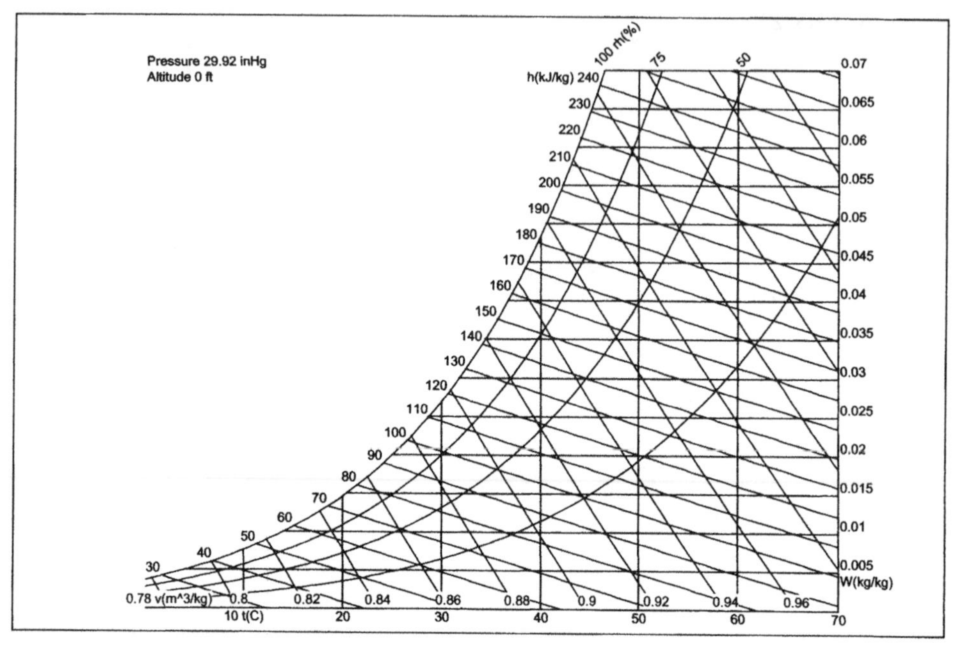

Figure A4.2: Low-temperature chart (Celsius degrees).

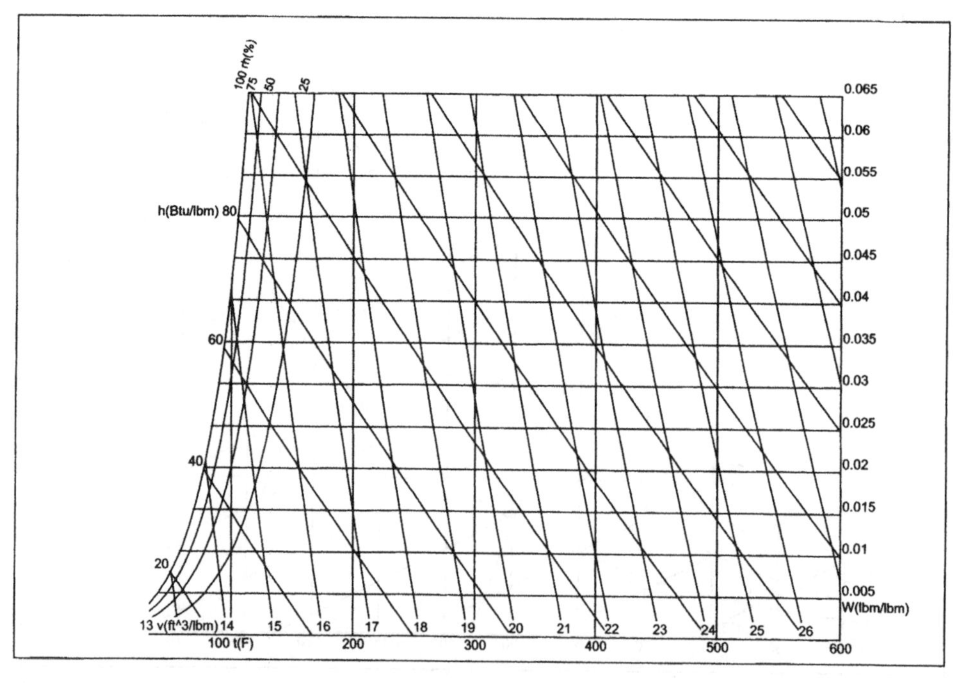

Figure A4.3: Intermediate-temperature chart (Fahrenheit degrees).

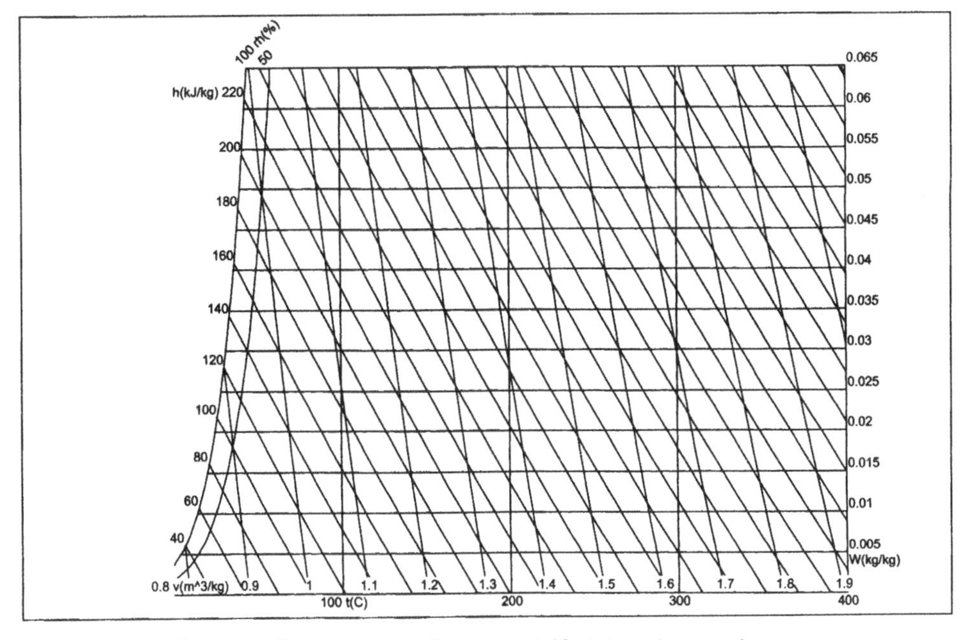

Figure A4.4: Intermediate-temperature chart (Celsius degrees).

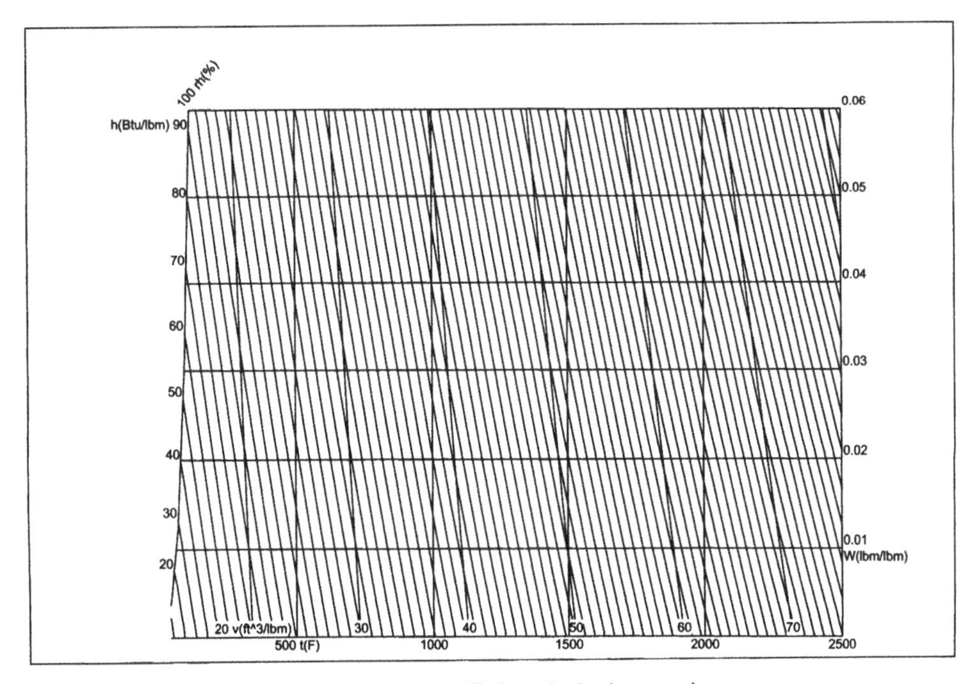

Figure A4.5: High-temperature chart (Fahrenheit degrees).

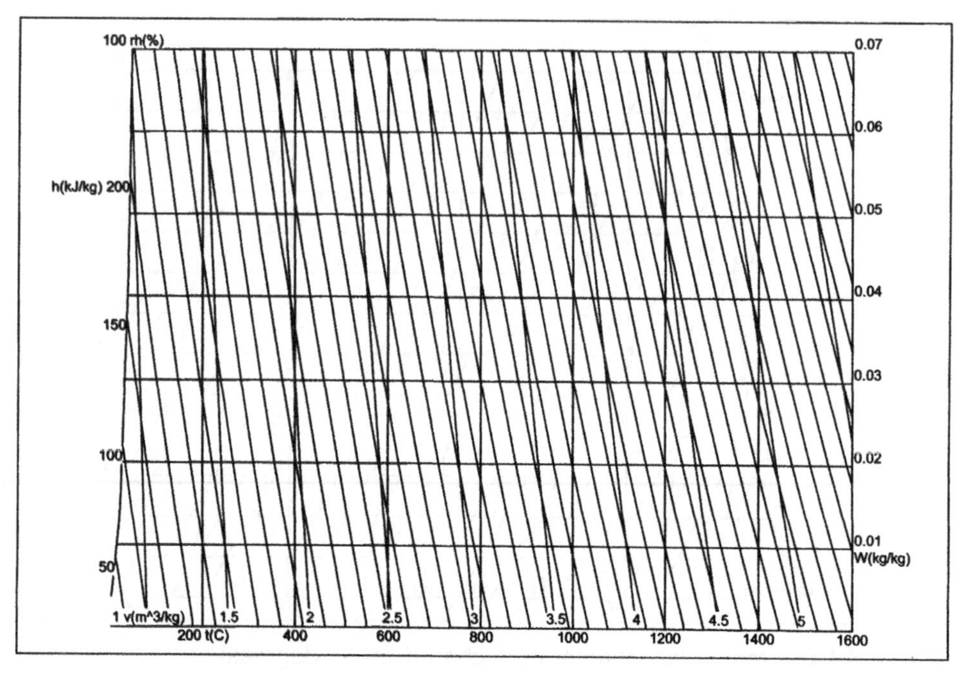

Figure A4.6: High-temperature chart (Celsius degrees).

Index

A

B

C

D

E

I

J

K

L

M

R

U

V

W

Printed and bound by CPI Group (UK) Ltd, Croydon, CR0 4YY

16/04/2025

14658442-0002